Micro World

The Science of the Invisible that Shapes the Visible

Micro World

The Science of the Invisible that Shapes the Visible

(Illustrated, Refined & Edited)

By: A.S. Raleigh and DD Editorials

DD BOOKS
Publishers for New Generations

DD BOOKS
Publishers for New Generations

Copyright Notice for Readers

Publication Year: **2026** | Publisher: **DD Books**

ISBN: 9789372446159

Edition: **First Edition, 2026** | Contact: ~~ddsalesglobal@gmail.com~~

Book Name: **Micro World**

Author: **A.S. Raleigh and DD Editorials**

Price: **Rs 250/-**

Table of Content

Reader Note 7

Publisher's Note 9

Introduction 11

Chapter-1. Microbiology - The Hidden Cornerstone of Life 13

Chapter-2. Timeline of Discovery - From Curiosity to Microbial Science 21

Chapter-3. Understanding Microorganisms? What they are and What they do 26

Chapter-4. Bacteria - Builders, Helpers, and Harmers 31

Chapter-5. Viruses - The Borderline Life Forms 35

Chapter-6. Fungi - Nature's Recyclers and Food Alchemists 40

Chapter-7. Protozoa - Single-Celled Swimmers of the Water World 45

Chapter-8. Algae - Microscopic Plants with Massive Power 50

Chapter-9. Where Microbes Live - From Deep Oceans to Our Skin 54

Chapter-10. Microbiomes - The Invisible World Inside us 59

Chapter–11. Microbes and Diseases - How Infections Happen 63

Chapter–12. Microbes in Food - From Kitchen Ferments to Factory Floors 68

Chapter–13. Microbes in Agriculture : Nature's Green Allies 73

Chapter–14. Microbes for a Cleaner Planet - Environmental Solutions 78

Chapter–15. Medical Microbiology - Healing with the Smallest Healers 83

Chapter–16. Microbes in Biotechnology Engineering the Invisible 88

Chapter–17. Space Microbiology - Life Beyond Earth 93

Chapter–18. Micro-Lab at School - DIY Experiments for Young Scientists 97

Chapter–19. Becoming a Microbe Detective - Projects and Inquiry 101

Chapter–20. Microbiology Careers - From Lab Coats to Space Suits 105

Reader Note

Every great journey begins with a moment of wonder—a quiet pause in which we realize that the world is far richer, deeper, and more mysterious than it first appears. Micro World: The Science of the Invisible that Shapes the Visible invites you into such a moment. This book is not merely a collection of scientific facts about microorganisms; it is a guided exploration into the hidden architecture of life itself. It reveals how countless unseen beings quietly build, maintain, disrupt, and renew the world we experience every day.

Within these pages, you will discover that the smallest forms of life often have the largest influence. From the microbes that make bread rise and yogurt tangy, to the bacteria that enrich soil and support crops, to the viruses that challenge medicine and reshape history, this book shows how invisible forces sculpt visible reality. Each chapter opens a new window into this astonishing microcosmos, carefully designed to spark curiosity, deepen understanding, and ignite imagination.

What makes this book special is its balance between intellectual depth and engaging storytelling. Complex scientific ideas are presented with clarity, warmth, and relevance. You do not need to be a scientist to appreciate its insights—only a willingness to ask questions and a desire to understand how life truly works beneath the surface. The chapters move from foundational concepts to real-world applications, guiding readers step by step through the hidden mechanics of microbiology. You will encounter microbes not as abstract textbook subjects, but as living participants in a grand planetary drama. They are builders and destroyers, healers and challengers, ancient survivors and futuristic tools. You will see how microbes live in the deepest oceans, inside glaciers, within volcanic vents, and even inside your own body—forming complex communities that influence digestion, immunity, mood, and overall health. The realization that you are never truly alone, that trillions of microscopic partners accompany you through life, can be both humbling and empowering.

This book also celebrates human curiosity. The timeline of discovery shows how generations of thinkers, tinkerers, and visionaries slowly unveiled the microbial world, often against disbelief and resistance. Their stories remind us that science advances through courage, persistence, and imagination. By understanding how knowledge was built, readers gain a deeper appreciation for the scientific process itself.

Beyond knowledge, this book offers possibility. You will see how microbes are harnessed to clean polluted environments, produce sustainable fuels, develop life-saving medicines, and even support space exploration. You will explore how biotechnology transforms invisible life into powerful tools for solving global challenges. For young readers and aspiring scientists, the later chapters provide hands-on inspiration—simple experiments, inquiry-based projects, and pathways toward careers that range from laboratories to outer space.

Publisher's Note

At the heart of every meaningful book lies a commitment to both truth and accessibility. As publishers, our mission is to honor the integrity of the author's vision while shaping the presentation in a way that speaks clearly and confidently to today's readers. Micro World: The Science of the Invisible that Shapes the Visible stands as a testament to that mission.

This book explores a subject that is foundational to modern science yet often misunderstood or overlooked. Microbiology affects nearly every aspect of human existence—from health and nutrition to agriculture, industry, and environmental sustainability. Our goal in bringing this work to readers is to ensure that its insights are not locked behind technical language or academic barriers, but instead flow with clarity, coherence, and purpose.

In preparing this edition, we have carefully preserved the author's central theme: that invisible life forms are not secondary characters in the story of Earth, but primary architects of its past, present, and future. At the same time, we have emphasized organization, readability, and narrative flow, allowing each chapter to build naturally upon the last. The result is a work that respects scientific rigor while remaining inviting to readers from a wide range of backgrounds.

We believe that modern readers seek more than information alone—they seek understanding, context, and relevance. This edition highlights how microbiology connects directly to everyday experiences: the food we eat, the illnesses we face, the technologies we rely on, and the environment we strive to protect. By weaving these connections throughout the text, we aim to make the science feel immediate and meaningful rather than distant or abstract.

Special attention has been given to maintaining a tone that is both authoritative and approachable. The voice of the book encourages curiosity rather than intimidation. It invites questions instead of assuming prior knowledge. This approach reflects our publishing philosophy: that education should empower, not exclude.

We also recognize the importance of inspiring the next generation of scientists and innovators. The inclusion of hands-on experiments, project ideas, and career pathways reflects our belief that books can serve as gateways to future possibilities. A single chapter, or even a single page, can spark a lifelong passion. We are proud to support works that carry this potential.

Every editorial decision behind this edition has been guided by respect—respect for the author's intent, respect for the scientific discipline, and respect for the reader's time and intelligence. We see Micro World not simply as a book, but as a bridge between curiosity and understanding, between the seen and the unseen.

It is our hope that this edition will serve students, educators, and general readers alike, offering both reliable knowledge and a sense of wonder. By bringing this work to a contemporary audience with clarity and care, we strive to ensure that its message continues to educate, inspire, and endure.

Introduction

There is a universe hidden within every drop of water, every breath of air, and every grain of soil. It is a universe alive with motion, struggle, cooperation, and transformation—a universe so small that it escapes our senses, yet so powerful that it shapes the destiny of all visible life. Micro World: The Science of the Invisible that Shapes the Visible invites you into this extraordinary realm. This book is a journey into the unseen foundations of existence. It explores how microscopic organisms, though invisible to the naked eye, govern processes that determine health and disease, growth and decay, balance and disruption. From ancient times to the frontiers of modern science, microbes have been silent companions to humanity, influencing our survival in ways we are only beginning to fully understand.

As you move through these chapters, you will encounter a world of astonishing diversity. You will meet bacteria that build ecosystems and others that cause devastation. You will explore viruses that blur the line between living and nonliving. You will discover fungi that recycle death into life and algae that quietly generate much of the oxygen we breathe. Each group of microorganisms carries its own story, its own strategies for survival, and its own role in the grand web of life. Yet this book is not only about microbes. It is also about people—about the scientists who dared to question what could not be seen, the innovators who developed tools to reveal the invisible, and the thinkers who connected microscopic processes to global patterns. Their curiosity and persistence form a narrative thread that runs throughout the book, reminding us that knowledge grows through imagination as much as through observation. A central theme of this work is interconnectedness. No organism exists in isolation. Microbes form vast networks within environments and within our own bodies, creating microbiomes that influence digestion, immunity, mental well-being, and overall vitality. Understanding these hidden partnerships reshapes how we think about ourselves—not as solitary beings, but as ecosystems in motion.

There is also an undercurrent of suspense and discovery woven into these pages. Each chapter uncovers another layer of complexity, another revelation about how life sustains itself against incredible odds. The more we learn, the more we realize how much remains unknown. This sense of mystery is not a weakness of science, but its greatest strength—it drives exploration, fuels innovation, and keeps the human spirit reaching forward.

Micro World bridges past, present, and future. It traces the historical roots of microbiology, examines its current applications, and looks toward emerging frontiers such as biotechnology, environmental restoration, and space exploration. In doing so, it presents microbiology not as a static body of knowledge, but as a living, evolving discipline. Above all, this book invites a shift in perspective. It asks you to slow down, to look closer, and to reconsider what truly matters in the architecture of life. The invisible is not empty. It is not insignificant. It is vibrant, dynamic, and essential. By opening this book, you step into a hidden dimension of reality—one that will change how you see the world around you and within you. The journey ahead is one of wonder, challenge, and profound insight. Welcome to the micro world.

Chapter-1.
Microbiology – The Hidden Cornerstone of Life

Step into a world where the smallest forms of life wield the greatest power. This chapter unveils the invisible forces shaping health, technology, climate, and the future of humanity, revealing why microbiology is the foundation of modern science and an essential key to understanding – and transforming – the world around us.

In a world dominated by visible marvels—skyscrapers, satellites, and smartphones—there exists an invisible realm more powerful, more ancient, and more influential than most realize. Microbiology is the science that unlocks this hidden universe. From the life-saving discoveries of antibiotics to the mysteries of deep-sea vents and the engineering of lab-grown meat, microbiology has shaped human progress at every turn. This chapter is designed to take youth on a fascinating journey through this unseen world, revealing why understanding microbes is not only crucial for careers in science and health, but also vital for solving 21st-century global challenges. Here, we explore what microbiology truly is, how it affects nearly every aspect of modern life, why it is becoming an essential subject for today's young generation, and how it weaves together the threads of multiple scientific disciplines—from medicine to outer space.

Microbiology is the branch of biology that deals with the study of microscopic organisms, commonly known as microbes.

These include bacteria, viruses, fungi, protozoa, and algae—organisms so small that they require a microscope to be seen. Although invisible to the naked eye, these tiny life forms are incredibly diverse and complex. They inhabit every corner of the planet—from boiling hot geysers and icy glaciers to the human gut and the soil beneath our feet. Some microbes cause diseases, while many others play critical roles in digestion, nutrient cycling, and even the production of food and medicine.

Microbiology isn't just about identifying microbes; it's about understanding their behavior, genetics, structure, interaction with the environment, and influence on all living systems. Whether you are curing an infection, creating vaccines, fermenting cheese, or treating wastewater, microbiology is at the heart of the process. It is a field that blends curiosity with purpose, offering a deep dive into the hidden biological forces that sustain and threaten life.

Microbiology's influence extends far beyond the walls of laboratories and petri dishes—it touches nearly every aspect of human life, making it one of the most dynamic and essential sciences of the modern era. In the field of health and medicine, microbes have completely transformed how we understand and treat diseases. The discovery of antibiotics such as penicillin, which emerged from the study of mold, marked the beginning of a medical revolution that continues to save millions of lives each year. Vaccines, another marvel rooted in microbial science, have eradicated or controlled devastating diseases like smallpox and polio. More recently, the intricate communities of gut microbes—collectively known as the gut microbiota—have captured the interest of researchers worldwide. These microscopic organisms are now recognized not only for their role in digestion but also for their surprising impact on mental health, immunity, and chronic conditions like obesity. The more we learn about these invisible allies and adversaries, the more central microbiology becomes to the future of personalized medicine and preventive healthcare.

Beyond the human body, microbiology plays a vital role in agriculture and food security. Nitrogen-fixing bacteria, for instance, convert atmospheric nitrogen into forms usable by plants, enriching soil fertility without the need for synthetic fertilizers. Certain fungi and microbial agents naturally suppress pests and plant diseases, reducing the reliance on harmful chemical pesticides. In the domain of food technology, microbes are the unsung heroes behind many of our favorite fermented foods such as yogurt, cheese, bread, and pickles. Not only do they enhance taste and texture, but they also improve nutritional value and shelf-life, making food safer and more sustainable. The science of fermentation, rooted in microbiology, is now experiencing a renaissance as researchers develop new, probiotic-rich foods to support better gut health.

Equally significant is the role of microbes in the climate and environment. Microorganisms are fundamental to the Earth's life-support systems. They drive the carbon and nitrogen cycles, break down organic waste, and maintain the ecological balance in oceans, soils, and the atmosphere. Certain microbes are capable of digesting toxic pollutants, including oil spills and industrial waste—a process known as bioremediation. Scientists are now engineering these tiny organisms to capture carbon dioxide, treat contaminated water, and reduce greenhouse gas emissions. These microbial solutions are becoming critical tools in the global fight against climate change, offering eco-friendly and cost-effective alternatives to conventional methods.

In the industrial and biotechnological sectors, microbes function as miniature powerhouses of production. Through fermentation and genetic engineering, they are used to manufacture life-saving medications such as insulin and human growth hormone, as well as essential enzymes for industries ranging from textiles to biofuels. Bacteria and yeast strains are routinely modified to produce biodegradable plastics, reducing our dependence on fossil-fuel-based products. One of the most groundbreaking applications of microbiology is in gene-editing technologies like CRISPR, which originated from bacterial immune systems. These tools are now enabling scientists to precisely alter DNA, holding the promise of curing genetic diseases and enhancing agricultural crops. Even space agencies are tapping into microbial potential—studying how microbes can be used to produce food, purify air and water, and recycle waste during long-term missions to Mars and beyond.

In every direction—health, agriculture, environment, and technology—microbiology offers transformative solutions. As the field rapidly evolves, it opens up a world of possibilities for young minds passionate about science, innovation, and sustainability. The invisible world of microbes may well hold the keys to solving some of the greatest challenges of our time.

In today's rapidly changing world, the need for a scientifically aware and socially responsible generation has never been more urgent—and microbiology stands out as one of the most essential sciences for empowering youth. The recent COVID-19 pandemic served as a powerful reminder of how a microscopic organism can bring global systems to a halt. It underscored the importance of understanding how viruses originate, spread, mutate, and can be contained. For young people, this knowledge isn't just for those pursuing careers in medicine or public health—it's critical for all future citizens. A generation educated in microbiology is far better prepared to respond to outbreaks, advocate for evidence-based policies, and participate in ef

Microbiology also sharpens the very core of scientific thinking. It encourages students to observe the invisible, ask questions, design experiments, and draw conclusions from evidence. These practices are not limited to laboratories—they build a foundation for critical thinking and scientific literacy that extends to every walk of life. Whether one becomes a science communicator, educator, journalist, policymaker, or designer, the ability to evaluate claims, test hypotheses, and understand biological processes makes for a more rational and informed worldview.

Moreover, microbiology is becoming a gateway to innovation and entrepreneurship. Around the globe, young scientists are using microbial knowledge to launch sustainable startups. From biodegradable packaging solutions and probiotic-based mental health supplements to microbes engineered for cleaning up oil spills, the possibilities are as endless as they are impactful. This intersection of creativity, ethics, and science is particularly appealing to today's socially conscious and tech-savvy youth, who want to create change and build meaningful careers.

Sustainability is another critical reason for youth to embrace microbiology. Understanding how microbes enrich soil, clean polluted water, and convert organic waste into energy is foundational to building greener communities. As cities grow and environmental challenges deepen, young people equipped with microbiological knowledge will lead the way in developing eco-friendly agriculture, urban composting programs, and renewable energy systems. Microbes, often seen as mere germs, are actually powerful agents of environmental healing and resilience.

By learning microbiology early, youth acquire a future-proof toolkit that spans biology, chemistry, environmental science, data analysis, and even engineering. It connects them to real-world issues and equips them with both the curiosity and capability to address them. In a century defined by global health, climate urgency, and technological innovation, microbiology isn't just another subject—it's a vital key to unlocking a better, smarter, and more sustainable future.

Microbiology is not a siloed discipline confined to the lab—it is a vibrant, interdisciplinary science that forms the backbone of several major fields shaping our future. Its influence radiates across medicine, environmental sustainability, agriculture, space exploration, and even cutting-edge technologies like artificial intelligence. This unique interconnectedness makes microbiology far more than a specialized branch of biology; it becomes a powerful gateway into the broader world of STEM and global innovation.

In the medical realm, microbiology is central to nearly every advancement in healthcare. Whether diagnosing infectious diseases, designing new antibiotics, or developing life-saving vaccines, medical professionals rely heavily on an understanding of microbes. Pathologists identify disease-causing organisms, immunologists explore how our immune systems respond to pathogens, and pharmacologists work to create microbe-targeting drugs. The rise of antibiotic resistance and emerging viral outbreaks makes microbiological literacy not just valuable—but indispensable to the future of medicine.

In the domain of environmental science, microbiology provides solutions for some of our most pressing ecological challenges. Environmental microbiologists investigate microbial communities in ecosystems like oceans, wetlands, forests, and even urban wastewater systems. These microscopic organisms play vital roles in recycling nutrients, breaking down pollutants, and restoring ecological balance. In fact, microbes are nature's original recyclers, quietly working to clean the planet long before humans ever attempted to do so. Microbial processes are now being harnessed to rehabilitate contaminated environments and to develop eco-friendly technologies for waste treatment and pollution control.

The frontiers of space exploration also increasingly depend on microbiology. As humanity sets its sights on Mars and deep-space missions, understanding how life can survive in extreme environments becomes essential. Microbiologists study extremophiles—organisms that thrive in boiling acid pools, frozen tundras, and high-radiation zones—to model the possibility of extraterrestrial life. These studies inform how we design space habitats, recycle resources in spacecraft, and prepare for long-duration missions. Microbes might someday help astronauts produce food, purify water, and manage waste in self-sustaining systems beyond Earth.

In agriculture and food technology, microbes are revolutionizing how we grow, preserve, and consume food. Beneficial bacteria are being used as biofertilizers to enhance soil fertility and reduce dependence on chemical fertilizers. Natural microbial pesticides are replacing synthetic chemicals, supporting healthier crops and ecosystems. Meanwhile, in the food industry, microbiology is enhancing safety, flavor, nutrition, and shelf life. From fermented foods to smart food packaging that senses spoilage through microbial changes, innovations in this field are helping feed a growing global population with less environmental strain.

Even in the highly technical world of data science and artificial intelligence, microbiology plays a key role. Modern microbial studies generate massive datasets, such as DNA sequences of microbial genomes or real-time tracking of disease outbreaks. Collaborations between microbiologists and data scientists are creating powerful tools that can model ecosystems, track epidemics, and predict microbial behavior under changing environmental conditions. Artificial intelligence is now being used to decode patterns in complex microbial networks—an approach that could lead to breakthroughs in personalized medicine and climate resilience.

This interdisciplinary nature of microbiology makes it one of the most exciting and relevant sciences for youth to explore. It breaks the boundaries between textbook knowledge and real-world application, offering multiple avenues for study, innovation, and career development. Whether a student dreams of becoming a doctor, environmentalist, engineer, astronaut, or data analyst, microbiology provides a versatile and deeply impactful foundation. It is not just a subject to study—it is a dynamic lens through which to understand and improve the world.

End of the Chapter

Chapter-2.
Timeline of Discovery – From Curiosity to Microbial Science

Travel through time to witness how simple curiosity unlocked an invisible universe. This chapter traces the bold ideas, accidental breakthroughs, and fearless scientists who transformed tiny observations into world-changing science, showing how each discovery built the foundation for modern medicine, technology, and our understanding of life itself.

The world of microbes has existed for billions of years, silently shaping the ecosystems of Earth and influencing the course of evolution itself. Yet it was only a few centuries ago that humanity first caught a glimpse of this hidden realm. The journey from mere curiosity to a full-fledged science of microbiology is one of the most remarkable chapters in human knowledge. This chapter unveils the timeline of major discoveries that transformed our understanding of life at the microscopic level—from the invention of the microscope and the revelation of unseen organisms to the groundbreaking innovations of genetic engineering and modern medicine. For today's youth, understanding this timeline is more than just history—it's a story of imagination, persistence, and paradigm shifts that continue to impact health, technology, food, and the environment. This chapter explores not only the major milestones but also the spirit of inquiry that drives microbiology forward, inspiring young minds to ask new questions and push the boundaries of what we know.

The journey of microbiology begins with an unusual but brilliant mind: Antonie van Leeuwenhoek. In the 17th century, using a handmade single-lens microscope that could magnify objects over 200 times, Leeuwenhoek became the first human in recorded history to observe and describe bacteria, protozoa, and other microorganisms. What he called "animalcules" were actually the microbial life forms that had existed unnoticed for millennia. His detailed letters to the Royal Society of London in the 1670s revealed a world no one had imagined. With this discovery, a completely new dimension of biology was born. Although Leeuwenhoek wasn't a trained scientist, his keen curiosity and craftsmanship laid the foundation for an entire field that would one day save millions of lives and unlock the secrets of life itself.

The key to Leeuwenhoek's breakthrough—and to all of microbiology—was the invention and evolution of the microscope. Though primitive magnifying lenses had been used for centuries, it was in the late 16th and early 17th centuries that compound microscopes were first developed. The Dutch spectacle makers Hans and Zacharias Janssen are often credited with the invention of the compound microscope around 1590, but it was scientists like Robert Hooke and Leeuwenhoek who truly brought its potential to life. Hooke's 1665 publication Micrographia described his observations of cork and coined the term "cell," which would become central to biology. As lenses improved over the centuries, scientists could observe microbes in more detail—identifying their structures, behaviors, and interactions with other organisms. The microscope was not just a tool; it was a portal into a universe that would redefine medicine, agriculture, and ecology.

While the ability to see microbes was an enormous leap, understanding their role in disease took nearly two centuries. For much of history, illness was believed to be caused by miasmas—bad air—or spiritual imbalance. This changed in the 19th century with the birth of the germ theory of disease. Scientists like Louis Pasteur and Robert Koch demonstrated through rigorous experimentation that microorganisms were not just invisible curiosities but agents of infection.

Pasteur's famous swan-neck flask experiment showed that microbes in the air could contaminate sterile solutions, debunking spontaneous generation theories. His development of the rabies and anthrax vaccines further cemented the medical importance of microbes. Around the same time, Koch identified the causative agents of diseases like tuberculosis and cholera, formulating "Koch's postulates" as a framework to link specific microbes to specific diseases. These breakthroughs not only advanced medicine but also set the stage for hygienic practices, hospital sanitation, and public health reforms around the world.

One of the most transformative milestones in microbiology came with the discovery and use of antibiotics. In 1928, Alexander Fleming accidentally discovered that a mold—Penicillium notatum—killed bacteria in his petri dish. This led to the development of penicillin, the first true antibiotic, which became widely available during World War II and revolutionized the treatment of bacterial infections. The antibiotic revolution saved countless lives and led to the search for other microbial products that could fight disease. At the same time, the concept of vaccination, pioneered earlier by Edward Jenner in the 18th century through smallpox inoculation, was evolving. Louis Pasteur's work on weakened pathogens expanded vaccination into a powerful preventive tool. Together, antibiotics and vaccines reshaped the medical landscape, turning once-deadly infections into manageable or preventable conditions.

As microbiology entered the 21st century, it began to intersect with digital technology and molecular biology in ways the early pioneers could never have imagined. Genetic engineering, particularly with the arrival of tools like CRISPR-Cas9, has enabled scientists to edit DNA with extraordinary precision. CRISPR, originally discovered as a bacterial immune defense mechanism, is now being used to correct genetic diseases, enhance crops, and design new treatments for cancer and viral infections. This breakthrough represents a fusion of microbiology with genomics, biotechnology, and artificial intelligence. Meanwhile, the global experience of pandemics, especially COVID-19, brought microbial science into everyday conversation. Understanding viral genomes, mutations, and transmission patterns became essential knowledge not just for scientists, but for the general public.

Another area where 21st-century microbiology is making waves is in phage therapy—the use of bacteriophages, viruses that infect and kill bacteria, to treat infections that no longer respond to antibiotics. With antibiotic resistance becoming a major global health threat, researchers are revisiting phages, which were once overshadowed by antibiotics, as potential lifesavers. Additionally, microbiome research—the study of microbial communities in our bodies and the environment—is revealing complex relationships between microbes and human health, including links to immunity, brain function, and chronic diseases.

The timeline of microbial discovery is not just a sequence of scientific facts; it is a living history of curiosity turned into capability. It reflects how the tiniest forms of life have transformed our understanding of the world and our ability to survive within it. For young learners, this journey offers not only inspiration but a reminder: the next great discovery could be waiting under a microscope, in a droplet of water, or hidden in the genetic code of a single cell. The story of microbiology is still being written—and its future chapters will likely be authored by the youth of today.

End of the Chapter

Chapter-3. Understanding Microorganisms? What they are and What they do

Meet the unseen architects of life. This chapter reveals who microorganisms truly are, how they survive, and why their tiny size hides enormous power—unlocking a deeper understanding of the invisible forces shaping health, ecosystems, technology, and the future of our planet.

To understand the powerful influence of microbiology, one must first become familiar with its central characters—the microorganisms themselves. These tiny life forms are among the most ancient and diverse beings on Earth. Though invisible to the naked eye, they outnumber all other organisms combined and shape nearly every ecosystem, from the human body to the deep ocean floor. This chapter serves as a gateway to identifying what microorganisms really are, how they behave, and how they differ from the plants and animals we see every day. For youth, understanding microorganisms is not just about memorizing types and traits—it's about uncovering a living world full of energy, adaptability, and mystery. Through this lens, young minds can begin to see microbes not as threats or abstract textbook terms, but as living, evolving forces that influence health, industry, climate, and even space exploration.

Microorganisms, or microbes, are organisms so small that most of them can only be seen under a microscope. Despite their size, they exhibit all the essential characteristics of life: they grow, reproduce, respond to stimuli, and adapt to their environment. These tiny entities can live in extreme conditions—from acidic hot springs and radioactive waste to the icy depths of Antarctica and the dry vacuum of space. Some microbes are single-celled, while others exist as colonies or simple multicellular organisms. What sets microorganisms apart is not just their invisibility, but their ability to evolve rapidly, colonize diverse environments, and perform complex biochemical tasks that larger organisms cannot. Their resilience and efficiency have made them the architects of evolution, the engines of ecosystems, and the inspiration for many modern scientific breakthroughs.

Microorganisms are broadly classified into several major groups, each with unique structures, functions, and ecological roles. The first group, bacteria, are single-celled prokaryotic organisms that lack a nucleus. They come in shapes like rods, spheres, and spirals, and can be found virtually everywhere—from soil and water to the human gut. While some bacteria cause diseases like tuberculosis or pneumonia, many are beneficial, aiding digestion, fixing nitrogen in the soil, or decomposing organic matter.

Viruses, on the other hand, are a biological paradox. They are not considered living in the traditional sense because they cannot reproduce on their own. Instead, they invade the cells of living organisms and hijack the host's machinery to replicate. Despite their simplicity—just genetic material wrapped in a protein coat—viruses have enormous impact. They cause diseases like influenza, HIV, and COVID-19, yet they also play roles in genetic research, evolution, and potential therapies like phage treatment.

Fungi are a fascinating group that includes yeasts, molds, and mushrooms. Unlike plants, fungi do not perform photosynthesis. Instead, they absorb nutrients from their surroundings, often by breaking down dead material. This makes fungi essential decomposers in ecosystems. Some fungi, like Saccharomyces cerevisiae, are used in baking and brewing, while others are sources of antibiotics, such as penicillin. However, some fungi can also cause diseases in plants, animals, and humans.

Protozoa are single-celled eukaryotic organisms, meaning they have a nucleus and complex internal structures. These microscopic creatures often live in water or moist environments and can move using hair-like cilia, whip-like flagella, or flowing extensions called pseudopodia. Protozoa play a vital role in nutrient cycles and food chains, particularly in aquatic ecosystems. Some, like Amoeba and Paramecium, are harmless or even beneficial, while others, such as Plasmodium—the parasite that causes malaria—can be deadly.

Algae, though often associated with pond scum or seaweed, are a diverse group of photosynthetic organisms that can range from microscopic phytoplankton to large kelp forests. Algae produce a significant portion of Earth's oxygen and serve as the base of aquatic food chains. While they resemble plants in their ability to photosynthesize, algae lack the complex structures of true plants such as roots, stems, and leaves. Some microscopic algae are also being explored for biofuel production and carbon capture technologies.

The way microorganisms live, move, and reproduce is a testament to the efficiency of life at its smallest scale. Many bacteria and protozoa move using appendages like flagella or cilia, while others glide or rely on the movement of fluids around them. Some microbes are stationary but incredibly efficient at colonizing new environments through spores or biofilms. Their reproduction varies by type: bacteria reproduce rapidly by binary fission, a process where one cell divides into two identical cells. Fungi often reproduce by releasing spores, which can travel long distances and remain dormant until favorable conditions arise. Viruses, lacking the machinery to reproduce on their own, inject their genetic material into host cells and turn them into virus-producing factories. This wide variety of survival and replication strategies gives microorganisms an unmatched adaptability in the natural world.

Unlike plants and animals, microorganisms operate on a scale and set of rules that make them uniquely powerful. Plants typically depend on sunlight and fixed root systems, while animals rely on mobility and sensory organs. Microorganisms, however, often live independently of light, oxygen, or fixed habitats. They absorb nutrients directly from their environment, and many can switch metabolic pathways to suit changing conditions. They can remain dormant for years and suddenly activate when the environment becomes favorable. This biochemical versatility and genetic flexibility make them entirely different from the more structured and visible life forms we interact with daily.

Understanding microorganisms also challenges the notion that size limits impact. Though microscopic, microbes can trigger pandemics, regenerate soil, produce life-saving drugs, and even shape planetary climates. For young learners, recognizing the sophistication of microbial life forms is a humbling and empowering step. It reveals that life operates on many levels—and that some of the most vital work is done out of sight. Through this awareness, youth not only gain scientific knowledge but also develop a broader respect for the hidden forces that sustain the living world.

End of the Chapter

Chapter-4. Bacteria – Builders, Helpers, and Harmers

Discover the double life of bacteria – tiny organisms that can both sustain and threaten life. This chapter uncovers how these microscopic builders power ecosystems, support human health, drive industry, and sometimes become dangerous foes, revealing why understanding bacteria is essential to shaping a safer, smarter future.

Among all the microbes that shape life on Earth, bacteria are perhaps the most misunderstood. Often associated with illness and infection, bacteria are in fact among the most ancient, diverse, and essential organisms in existence. They inhabit virtually every environment – from volcanic vents deep beneath the ocean to the inside of the human body – and play roles both beneficial and harmful. This chapter invites youth to look beyond the common fear of "germs" and explore the fascinating world of bacteria. We will examine their shapes, how they multiply at astonishing speed, how they serve as silent allies in everything from digestion to industry, and how some can also become dangerous pathogens that challenge global health. Understanding bacteria not only provides insights into biology and medicine, but also helps young minds appreciate the complex balance between living organisms and the ecosystems they sustain or disrupt.

Bacteria are unicellular prokaryotes, meaning they lack a nucleus and other membrane-bound organelles. Despite their simplicity, they are highly organized and efficient life forms. One of the first ways scientists identify bacteria is by observing their shape. The most common bacterial shapes include cocci (spherical), bacilli (rod-shaped), and spirilla (spiral or corkscrew-shaped). These shapes are more than just appearances—they influence how bacteria move, reproduce, and survive in different environments. Cocci, for instance, often appear in clusters or chains and are common in infections like strep throat. Bacilli are well-suited for nutrient absorption due to their elongated form, while spirilla can move rapidly through viscous fluids, which helps them thrive in places like the human digestive tract or contaminated water.

One of the most remarkable features of bacteria is their ability to multiply rapidly. They reproduce through a simple process called binary fission, where one bacterial cell duplicates its DNA and divides into two identical daughter cells. Under ideal conditions—such as warmth, moisture, and plenty of nutrients—some bacteria can double in number every 20 minutes. This means a single bacterium could give rise to millions of descendants in just a few hours. This rapid reproduction enables bacteria to colonize environments quickly, adapt to new conditions, and, in the case of harmful strains, overwhelm a host's immune system. However, this same quality also makes bacteria ideal for scientific research and industrial applications, as large quantities can be grown swiftly and cost-effectively.

Not all bacteria are harmful; in fact, many are indispensable to life. In the human body, especially in the digestive system, beneficial bacteria help break down complex food particles, synthesize certain vitamins, and support immune function. These helpful bacteria, collectively known as the gut microbiota, form a symbiotic relationship with their host, influencing everything from metabolism to mood. In the soil, bacteria play an essential role in nutrient cycling. Nitrogen-fixing bacteria convert atmospheric nitrogen into forms plants can absorb, boosting soil fertility without artificial fertilizers. In the food industry, bacteria are used in the fermentation of yogurt, cheese, vinegar, and pickles—processes that enhance flavor, texture, and nutritional content. Industrially, bacteria help produce enzymes, biofuels, and biodegradable plastics, making them vital to sustainable technologies and green innovation.

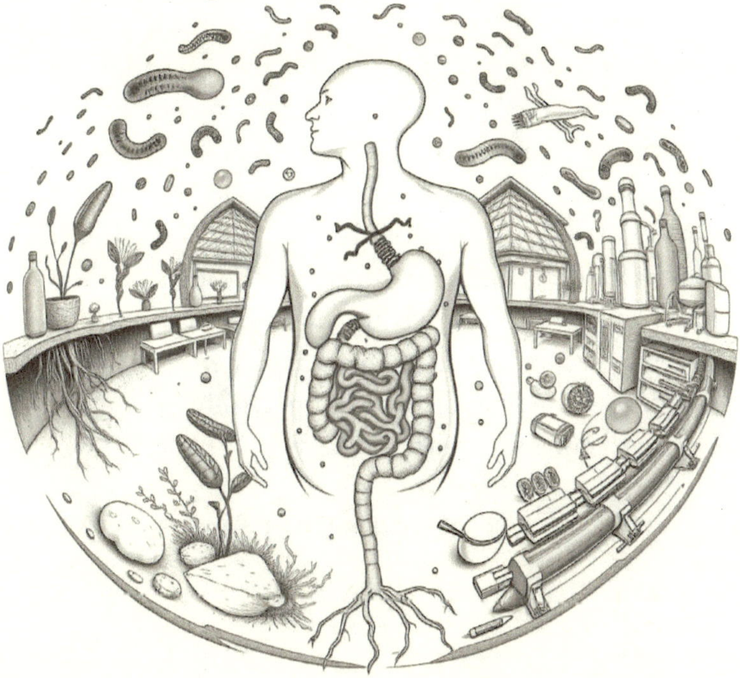

Despite their benefits, some bacteria are harmful and cause a wide range of diseases in humans, animals, and plants. These pathogenic bacteria invade the body, multiply, and produce toxins or trigger immune responses that lead to illness. Diseases like tuberculosis, cholera, pneumonia, and food poisoning have bacterial origins. The rise of such diseases has led to the development and widespread use of antibiotics—drugs designed to kill or inhibit bacterial growth. However, overuse and misuse of antibiotics have led to one of the greatest modern health threats: antibiotic resistance. When bacteria evolve to survive antibiotic treatment, they become "superbugs" that are difficult or sometimes impossible to treat. This challenge is compounded by the slow pace of developing new antibiotics and the global spread of resistant strains. Antibiotic resistance not only endangers individual patients but also threatens to undo decades of medical progress, making once-treatable infections potentially deadly.

In today's world, understanding bacteria is more important than ever. They are both friends and foes, capable of nurturing ecosystems or causing epidemics. For youth, learning about bacteria provides essential insights into biology, health, ecology, and biotechnology. It also fosters critical thinking about how human actions—such as medical practices, agricultural choices, and environmental policies—impact microbial life and, in turn, our own survival. Bacteria are builders of health, helpers in nature and industry, and harmers when misunderstood or mishandled. To navigate the future wisely, we must learn to work with them, not against them.

End of the Chapter

Chapter-5.
Viruses – The Borderline Life Forms

Enter the strange borderland between life and non-life, where viruses quietly shape history. This chapter uncovers how these tiny invaders operate, evolve, and challenge humanity, revealing why understanding viruses is essential for protecting health and preparing for future global threats.

Viruses are among the most mysterious and controversial entities in the biological world. Invisible to the naked eye and far simpler than even the smallest bacteria, viruses blur the line between living and non-living things. They cannot eat, grow, or reproduce on their own—yet they have the power to infect, evolve, and sometimes devastate entire populations. This chapter offers young readers a captivating exploration into the world of viruses, revealing how these microscopic agents work, why they behave differently from all other life forms, and how science continues to battle against the most dangerous viral threats. From the common cold to global pandemics, viruses have a profound influence on human health, and understanding them is essential not only for future doctors and scientists but for every informed individual. As youth delve into the structure, behavior, and societal impact of viruses, they will discover that these tiny particles carry both danger and scientific wonder.

The question of whether viruses are truly alive remains one of the great debates in biology. Unlike bacteria or other microbes, viruses do not consist of cells, do not perform metabolism, and cannot reproduce independently. Outside of a living host, a virus is simply a small packet of genetic material encased in a protein shell—essentially inert. Yet once it enters a living cell, it springs to action, hijacking the host's biological machinery to make copies of itself. This dual nature is what places viruses in a gray zone: they are lifeless outside the body, yet astonishingly active inside. Scientists often refer to viruses as "replicators" rather than living organisms, emphasizing their dependency on host life to function. Despite their simplicity, viruses are far from primitive. They have evolved clever strategies to invade hosts, evade immune systems, and survive across generations.

The structure of viruses is both elegant and efficient. Most viruses are made up of genetic material—either DNA or RNA—enclosed in a protein coat called a capsid. Some viruses also have an outer envelope derived from the host cell's membrane, which helps them infect other cells. The shape of the virus varies widely: some are spherical, others are rod-shaped, and some, like bacteriophages, have a spider-like appearance. What all viruses share is a lack of cellular components—they have no nucleus, organelles, or cytoplasm. When it comes to reproduction, viruses do not divide like cells. Instead, they insert their genetic material into a host cell and take over its machinery to produce viral proteins and replicate their genome. The host cell, overwhelmed by this hijacking, eventually bursts or releases new viruses to infect neighboring cells. This process can be rapid and destructive, which is why many viral infections spread quickly and cause acute illness.

One of the most fascinating and concerning traits of viruses is their ability to mutate. Since many viruses—especially RNA viruses—replicate with low accuracy, they accumulate mutations frequently. Some of these changes are harmless, but others can help viruses evade the immune system, become more contagious, or resist antiviral drugs. This is why new strains of the flu appear every year and why the coronavirus that causes COVID-19 evolved into multiple variants. The interaction between viruses and their hosts is a complex arms race. While the immune system tries to recognize and destroy viruses, the viruses adapt in return, finding new ways to survive. Some viruses, like HIV, can hide inside the host's genome for years, while others like the common cold quickly replicate and spread before the immune system can mount a full defense. These strategies make viruses highly adaptable and difficult to eliminate entirely.

Throughout history, viruses have been responsible for some of the most significant diseases in human society. The common cold, caused by various rhinoviruses and coronaviruses, is a mild yet persistent reminder of viral presence. Influenza has triggered global pandemics, such as the deadly 1918 Spanish flu. HIV, the virus that causes AIDS, has claimed millions of lives and remains a major public health concern. More recently, the COVID-19 pandemic caused by the novel coronavirus SARS-CoV-2 changed the world almost overnight—disrupting economies, altering lifestyles, and showcasing the urgent need for global cooperation in the face of viral threats. These examples illustrate not only the medical burden of viruses but also their ability to influence politics, economics, and even social behavior. They remind us that viral diseases are not just biological issues—they are deeply intertwined with the human experience.

In the face of such challenges, one of humanity's most powerful tools is the development of vaccines. Vaccines work by training the immune system to recognize a virus before a real infection occurs. By introducing a harmless piece of the virus—such as an inactivated form, a protein, or genetic material—the body learns to mount a defense. Vaccines have been instrumental in eradicating smallpox, controlling polio, and drastically reducing measles and hepatitis infections. The rapid development of mRNA vaccines during the COVID-19 pandemic marked a new era in vaccine science, offering flexible platforms to respond quickly to emerging diseases. Beyond vaccines, preventive strategies like sanitation, hygiene, vector control, and public awareness campaigns are critical in limiting viral spread. However, vaccine hesitancy and unequal access remain major obstacles, particularly in low-income regions. Educating young minds about the science and significance of vaccination is key to building a healthier future.

Viruses, though simple in structure, are deeply complex in impact. They challenge our understanding of life, stretch the boundaries of biology, and test the resilience of our healthcare systems. Yet they also inspire breakthroughs in science—from genetic research to innovative therapies. For the youth of today, exploring the world of viruses is more than an academic exercise—it is a journey into one of the most dynamic and urgent frontiers of science. By learning how viruses operate, evolve, and can be controlled, young people become better equipped to participate in the global mission of health, preparedness, and innovation. In this microscopic battlefield between host and invader, knowledge itself becomes the most powerful form of protection.

End of the Chapter

Chapter-6.
Fungi – Nature's Recyclers and Food Alchemists

Step into the hidden kingdom that turns decay into life and simple ingredients into wonders. This chapter reveals how fungi recycle nature's waste, craft our favorite foods, and fuel medical breakthroughs – showing why these quiet alchemists are essential to Earth's balance and humanity's future.

Fungi are one of nature's most overlooked but most powerful life forms. Neither plant nor animal, fungi occupy a unique kingdom of life that is essential to the survival of ecosystems, the richness of our diets, and even the innovation of modern medicine. From the soft yeasts that make bread rise to the towering mushrooms in forests, fungi are nature's great recyclers and hidden chemists. This chapter will take youth deep into the often-invisible world of fungi, showing how these organisms break down dead matter, enrich the soil, create delicious and life-saving products, and, when unchecked, also cause serious diseases. Through an understanding of fungi, young minds will uncover the incredible ways life renews itself, the role of decay in creation, and the microscopic partnerships that keep our planet alive and flourishing.

Fungi come in many shapes and forms, but they all share some key characteristics. Unlike plants, they do not perform photosynthesis. Instead, fungi absorb nutrients from organic material around them, often through intricate networks of thread-like structures called hyphae. These hyphae form a web called mycelium, which stretches through soil, wood, or other materials like a hidden internet of the natural world. Among the main types of fungi are yeasts, molds, and mushrooms—each serving distinct roles in nature and industry. Yeasts are single-celled fungi best known for their role in fermentation. They are tiny yet mighty, converting sugars into alcohol and carbon dioxide, making them essential for baking, brewing, and winemaking. Molds are multicellular fungi with fuzzy or filamentous growth; while some molds spoil food, others, like Penicillium, produce antibiotics and fine cheeses. Mushrooms, the fruiting bodies of certain fungi, emerge above ground as the reproductive structures of mycelial networks, releasing spores to start new colonies. These familiar caps and stalks are just a small visible part of a vast fungal system working tirelessly below the surface.

One of the most vital roles fungi play in ecosystems is decomposition. They are nature's recyclers—breaking down dead plants, animals, and organic waste into simpler compounds that return nutrients to the soil. Without fungi, forests would choke on their own fallen leaves and dead wood, and nutrient cycles would stall. Fungi decompose complex substances like lignin and cellulose, which few other organisms can digest. In doing so, they release essential elements such as nitrogen, phosphorus, and carbon back into the environment, fueling new growth and sustaining biodiversity. Some fungi also form symbiotic relationships with plant roots, known as mycorrhizae. These partnerships enhance water and nutrient absorption for the plant while feeding the fungus with carbohydrates. This cooperative alliance is so successful that over 90% of plant species depend on it for healthy growth. Through such unseen teamwork, fungi quietly maintain balance and fertility in Earth's ecosystems. Beyond their ecological importance, fungi are responsible for some of the most magical transformations in our food culture.

In kitchens and factories around the world, fungi serve as food alchemists—turning raw ingredients into flavorful, nutritious, and sometimes intoxicating creations.

Yeast, especially Saccharomyces cerevisiae, is indispensable in baking. It feeds on sugars in dough, producing carbon dioxide that makes bread rise and gives it a soft, airy texture. In brewing and winemaking, the same yeast ferments sugars into alcohol, creating beer, wine, and spirits. Molds like Penicillium roqueforti and Penicillium camemberti help ripen blue cheeses and soft cheeses such as Brie and Camembert, adding unique tastes and textures that define gourmet cuisine. Fungi are also being explored for meat alternatives, such as fungal protein products that offer sustainable, high-protein food with a much lower environmental footprint. These culinary contributions are not only delicious but demonstrate fungi's versatility in meeting global food challenges through innovation and tradition.

However, not all interactions with fungi are beneficial. Some fungi are capable of causing diseases in both plants and humans, often with serious consequences.

In humans, fungal infections can range from mild conditions like athlete's foot and yeast infections to more severe illnesses such as candidiasis, aspergillosis, or cryptococcal meningitis—especially in individuals with weakened immune systems. These diseases can be difficult to treat, as fungal cells share many biological features with human cells, making it challenging to target them without harming the host.

In agriculture, fungal pathogens devastate crops and reduce food security worldwide. Fungi like Puccinia (rust), Fusarium, and Botrytis cause blights and rots that destroy harvests and damage economies. Some fungi also produce mycotoxins—harmful substances that contaminate grains, nuts, and stored foods, posing risks to both human and animal health. These dangers make fungal research crucial for protecting both public health and global food supplies.

Fungi are an ancient, intelligent, and indispensable force in the natural world. They clean up waste, feed plants, support our diets, and even save lives through medicines like penicillin and immunosuppressants.

Yet they also remind us of the thin line between harmony and imbalance, between nourishment and danger.

For youth exploring science, fungi offer an extraordinary example of how complexity, cooperation, and hidden beauty exist at every scale of life. From forest floors to fermentation tanks, fungi reveal that nature thrives on transformation—and that even in decay, there is opportunity for creation.

Understanding fungi isn't just about knowing another group of organisms—it's about learning how life sustains itself through cycles of giving, taking, and renewal.

End of the Chapter

Chapter-7.
Protozoa – Single-Celled Swimmers of the Water World

Dive into the lively waters where single-celled explorers swim, hunt, and survive. This chapter reveals how protozoa shape aquatic ecosystems, influence human health, and demonstrate the remarkable complexity hidden within just one living cell.

Beneath the surface of ponds, rivers, oceans, and even tiny droplets of water, lives a bustling, unseen world of organisms that are among the most diverse and dynamic creatures on Earth – protozoa. These single-celled eukaryotes are like microscopic adventurers, equipped with the ability to move, hunt, escape, and adapt to rapidly changing aquatic environments. Though small, they play massive roles in ecosystems and human health. This chapter dives deep into the fascinating biology of protozoa, revealing how their unique movements, structures, and behaviors shape the aquatic food web and influence global health through the diseases they transmit. For youth, understanding protozoa provides not just a new appreciation for life at the micro-level, but also insight into the complexity of evolution, adaptation, and the delicate interconnection between all living things.

Protozoa are single-celled organisms that belong to the kingdom Protista. Unlike bacteria, protozoa are eukaryotic, which means their cells contain a true nucleus and other membrane-bound organelles. This cellular complexity allows them to perform many of the same functions as multicellular animals—despite being just one cell.

Some protozoa are free-living, feeding on bacteria and other microscopic organisms, while others are parasitic and depend on host organisms for survival.

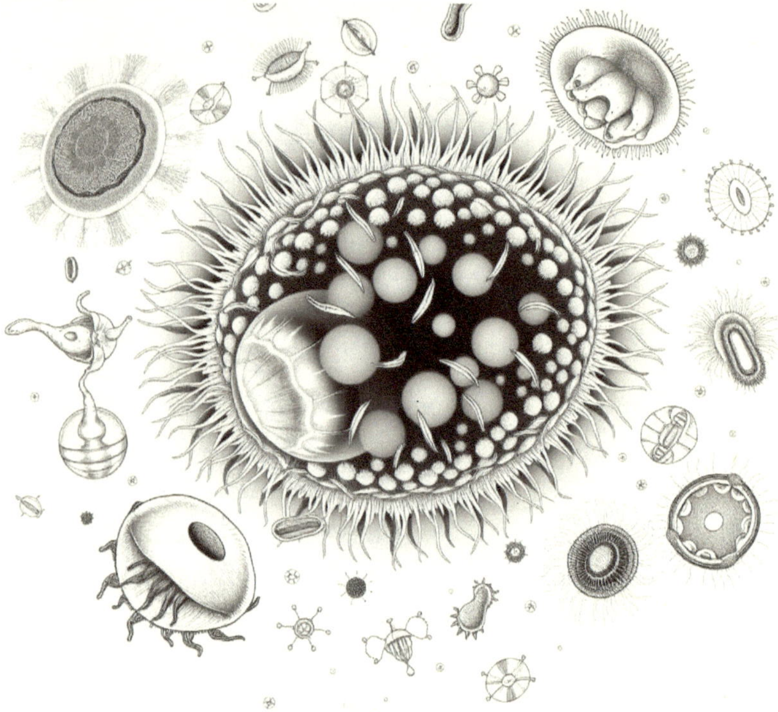

Their size usually ranges from a few micrometers to a few millimeters, but their structural diversity is enormous. Many protozoa have flexible cell membranes, allowing them to change shape, while others possess protective outer layers or even internal skeletons.

They can respond to environmental stimuli, engulf food through a process called phagocytosis, and reproduce either sexually or asexually. These qualities make protozoa more than just microscopic life—they are models of survival, resilience, and biological creativity.

What makes protozoa particularly fascinating is their varied and often graceful methods of movement. These organisms are not passive drifters; they are active swimmers, crawlers, and explorers of their microscopic world. One major form of movement is through cilia—tiny hair-like projections that beat in coordinated waves to propel the organism forward or sweep food into its mouth-like opening. Paramecium, a slipper-shaped protozoan, is a classic example of a ciliate, using thousands of these hairs for both motion and feeding.

Another method of movement involves flagella—long, whip-like tails that spin or wave to drive the organism through water. Euglena, a flagellated protozoan, can swim quickly using its single flagellum and even photosynthesize like a plant, making it a hybrid of animal- and plant-like behaviors.

Then there are amoeboid protozoa, like Amoeba proteus, which move by extending parts of their body called pseudopodia, or "false feet." These flowing extensions help them creep along surfaces and engulf food particles in a slow, fluid dance. Each method of movement not only reflects how protozoa navigate their environments but also demonstrates the incredible adaptability of life—even in its simplest forms.

Despite their often benign presence in aquatic ecosystems, some protozoa are responsible for serious and sometimes deadly diseases in humans. One of the most well-known protozoan diseases is malaria, caused by the Plasmodium parasite. Transmitted by the bite of female Anopheles mosquitoes, malaria affects hundreds of millions of people every year and has been a major cause of mortality throughout human history. Once inside the human body, Plasmodium undergoes complex life cycle stages in the liver and red blood cells, leading to fever, chills, and, if untreated, severe complications or death. Another protozoan disease is amoebic dysentery, caused by Entamoeba histolytica, which spreads through contaminated food or water and leads to gastrointestinal distress, diarrhea, and abdominal pain. There are also other protozoa like Trypanosoma (causing African sleeping sickness) and Giardia (causing giardiasis), which have major public health implications, particularly in regions with limited access to clean water and healthcare. The study and control of these diseases highlight the importance of protozoa in medical science and public health strategy.

Beyond their disease-causing potential, protozoa are vital players in aquatic ecosystems and are essential to the health of freshwater and marine environments. At the base of the aquatic food chain, many protozoa feed on bacteria and microscopic algae, keeping microbial populations balanced and preventing overgrowth that could harm water quality. In turn, protozoa themselves serve as a food source for small invertebrates and larvae, which are eaten by larger aquatic animals, supporting the entire ecosystem. Some protozoa also participate in nutrient cycling, breaking down organic matter and helping to release elements like nitrogen and phosphorus back into the environment. In coral reefs and other symbiotic systems, protozoa such as zooxanthellae (technically algae but often grouped with protozoan-like symbionts) live inside corals and provide them with energy through photosynthesis, maintaining the vibrant colors and health of these ecosystems. Even in the harshest environments—like deep-sea vents or frozen polar lakes—protozoa have been found thriving, proving their incredible adaptability and ecological importance.

Protozoa, though often overlooked, represent a stunning example of life's ingenuity. They embody complexity within simplicity, intelligence without brains, and influence without visibility. For youth, exploring the world of protozoa can be a source of wonder and empowerment. These tiny swimmers reveal that the greatest forces in nature are not always the most visible. They teach us that motion, adaptation, survival, and even danger can come in microscopic packages—and that by understanding these hidden creatures, we gain clearer insight into ourselves, our ecosystems, and the intricate web of life that binds us all.

End of the Chapter

Chapter-8.
Algae – Microscopic Plants with Massive Power

Discover how tiny green life forms quietly power the planet. This chapter reveals how algae generate oxygen, fuel food webs, and inspire sustainable energy solutions—proving that some of Earth's greatest forces begin at the smallest scale.

Algae might seem like mere green scum on a pond or strands of seaweed swaying beneath the waves, but these organisms are among the most powerful and productive life forms on Earth. Often ignored or underestimated, algae are responsible for producing more than half of the planet's oxygen and are the unsung engines of aquatic ecosystems. From single-celled phytoplankton to sprawling kelp forests, algae exist in incredible variety and are at the center of emerging solutions for food, fuel, and climate sustainability. This chapter explores the inner workings of algae, how they differ in form and function, how they support life as primary producers, and how their unchecked growth can sometimes cause ecological harm. For young readers, algae offer an inspiring glimpse into how even the smallest life forms can hold immense influence over global systems—and the future of human innovation.

Algae are remarkable organisms capable of performing photosynthesis, the process by which light energy from the sun is converted into chemical energy to produce food. Unlike most plants, algae do not have roots, stems, or leaves, but they contain chlorophyll and other pigments that allow them to absorb sunlight. Through photosynthesis, algae release vast amounts of oxygen into the atmosphere—making them critical not only for aquatic life but for all life on Earth. In fact, microscopic algae in oceans, particularly phytoplankton, are responsible for producing at least 50–60% of the planet's oxygen. These tiny floating cells are like living solar panels, capturing sunlight and using it to drive the most fundamental energy process on the planet. In doing so, they form the very foundation of aquatic food webs while playing a vital role in regulating Earth's climate by absorbing large amounts of carbon dioxide.

Algae exist in an incredible range of forms—from single-celled microscopic organisms to large, multicellular seaweeds that stretch for meters underwater. Unicellular algae, such as diatoms and dinoflagellates, are often found in freshwater and marine environments and drift freely as part of the plankton community. Despite their size, they are enormously productive and reproduce rapidly, especially in nutrient-rich waters. These single-celled organisms possess intricate cell walls made of silica or cellulose, and some even glow in the dark due to bioluminescence.

On the other end of the spectrum are multicellular algae like red algae, brown algae (such as kelp), and green algae, which can form large colonies or complex structures.

Kelp forests, for instance, provide shelter, food, and breeding grounds for countless marine animals, making them underwater equivalents of terrestrial rainforests. While both unicellular and multicellular algae share the ability to photosynthesize, their ecological roles and physical complexity vary widely, showcasing the immense diversity of life within this group.

In aquatic food chains, algae serve as the primary producers—organisms that create their own food and serve as the base of the food web. Phytoplankton are eaten by zooplankton, which are then consumed by small fish, which in turn are prey for larger fish, birds, and marine mammals. Without algae, the entire marine food web would collapse. But algae are more than just a food source for fish—they're becoming a promising resource for humans, too.

Certain algae can be harvested and processed into biofuels—renewable energy sources derived from organic material. Algal biofuels have the potential to reduce our reliance on fossil fuels because they grow quickly, don't require arable land, and can absorb carbon dioxide while producing oxygen. In addition, edible algae like spirulina and chlorella are being promoted as superfoods rich in proteins, vitamins, and antioxidants. As we look for sustainable ways to feed and power the world, algae stand out as tiny organisms with massive potential to support human life while protecting the planet.

Yet, not all algal activity is beneficial. Under certain conditions—such as excessive nutrients from agricultural runoff or warm water temperatures—algae can grow uncontrollably in what are called algal blooms.

These rapid increases in algal populations, particularly of toxin-producing species, can disrupt aquatic ecosystems, block sunlight from reaching underwater plants, and deplete oxygen levels when the algae die and decompose.

This oxygen depletion creates "dead zones" where few aquatic organisms can survive. Harmful algal blooms (HABs), such as red tides caused by dinoflagellates, can also produce toxins that poison fish, shellfish, and even humans who consume contaminated seafood.

These events pose serious threats to fisheries, tourism, and water safety. While algae are crucial for ecological balance, their unchecked growth reveals how human impact—through pollution, climate change, and habitat alteration—can turn even beneficial organisms into sources of environmental stress. Algae, in all their diversity and power, represent the awe-inspiring adaptability of nature.

They are ancient, efficient, and vital—acting as oxygen producers, carbon sinks, food sources, and biofactories all at once. For young learners, studying algae is an opportunity to appreciate how interconnected life is, from the smallest plankton to the largest whale, and how our choices affect the invisible systems that keep Earth alive.

Algae challenge us to think differently about power—not just in terms of strength or size, but in the quiet persistence of sunlight, cells, and cycles. Whether in a drop of water or a fuel tank of the future, algae remind us that true strength often lies in the overlooked, the unseen, and the seemingly small.

End of the Chapter

Chapter-9.
Where Microbes Live – From Deep Oceans to Our Skin

Journey across Earth's most extreme landscapes and into your own body to discover where microbes thrive. This chapter reveals how invisible life colonizes every environment imaginable, shaping ecosystems, sustaining health, and redefining what it truly means for life to survive.

Microbes are everywhere—not just in the obvious places like muddy ponds or moldy bread, but in the most unexpected corners of the world and even on and inside our bodies. These microscopic life forms have colonized every environment imaginable, from the oxygen-rich soil beneath our feet to the toxic sludge of radioactive waste. They live in boiling geysers, frozen glaciers, the depths of the ocean, and the thin air of the stratosphere. This chapter explores the astonishing variety of habitats microbes call home and the roles they play in shaping ecosystems. For today's youth, discovering where microbes live isn't just a scientific curiosity—it's a journey into the resilience and versatility of life itself. By understanding their habitats, we also understand their impact on the Earth's systems and our own bodies, revealing a hidden world that constantly supports, transforms, and challenges our everyday existence.

Microorganisms have evolved to thrive in nearly every natural environment. In soil, microbes like bacteria and fungi are responsible for decomposing organic matter, recycling nutrients, and forming the foundation of the underground ecosystem.

They work in close association with plant roots, enabling the uptake of essential minerals like nitrogen and phosphorus. Without these soil microbes, agriculture as we know it would collapse, and natural plant growth would falter.

In water environments—both freshwater and marine—microbes such as algae, protozoa, and planktonic bacteria float or swim in the upper layers, driving photosynthesis and oxygen production. Others cling to rocks, sediments, or aquatic plants, digesting waste and forming the basis of aquatic food chains. Even the air teems with microbial life.

Tiny spores, pollen, and bacterial cells are carried on wind currents, often traveling hundreds of miles. Meanwhile, our own bodies host rich microbial communities. The human skin, mouth, gut, and even nostrils provide diverse microhabitats for thousands of microbial species, many of which are beneficial allies that help with digestion, immunity, and disease prevention.

The diversity of microbial life is staggering. Scientists estimate that there are more microbial species on Earth than stars in the galaxy—most of which have yet to be discovered or fully studied. This microbial biodiversity is essential for ecosystem balance and function.

Microbes regulate nutrient cycles, purify water, detoxify pollutants, and support higher life forms. In forests, microbial decomposers return carbon and nitrogen to the soil. In oceans, planktonic microbes help trap carbon dioxide from the atmosphere and release oxygen back into the air.

In deserts, microbes form biological soil crusts that prevent erosion and retain moisture. The more diverse the microbial community, the more resilient an ecosystem is to disturbances like climate change, disease outbreaks, and pollution. Microbial biodiversity is not just a measure of quantity—it's a measure of stability, adaptability, and ecological intelligence. Preserving this hidden biodiversity is as important as protecting forests and wildlife, for it forms the living web that sustains all life.

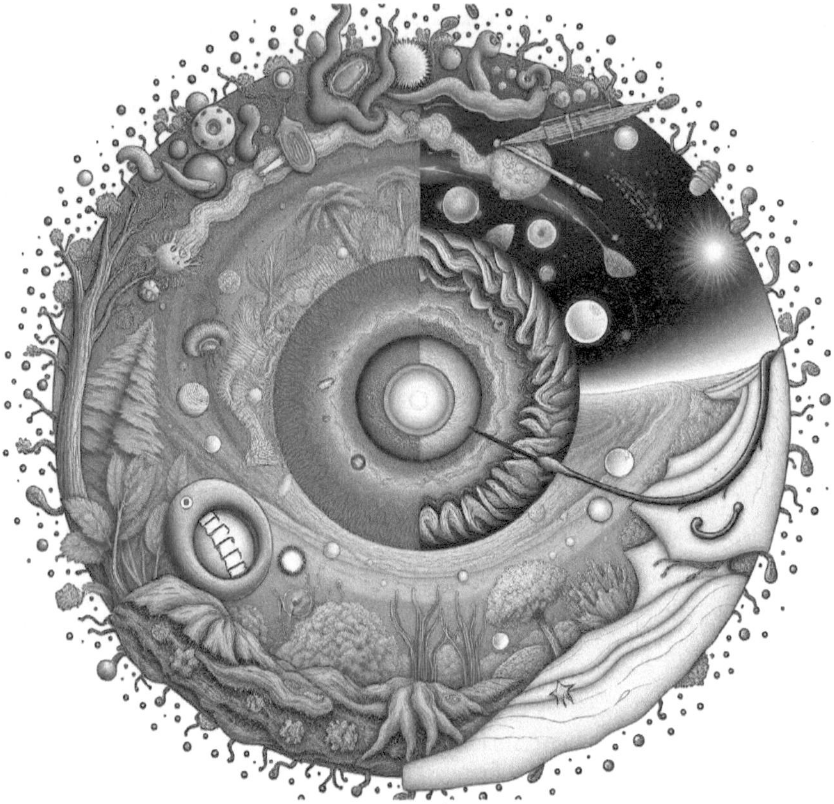

Among the most mind-blowing aspects of microbial life is the existence of extremophiles—microbes that not only survive but thrive in extreme environments that would kill most other organisms. These include thermophiles that live in boiling hot springs or hydrothermal vents, psychrophiles that thrive in glaciers and polar ice caps, halophiles that love salty lakes like the Dead Sea, and acidophiles that flourish in environments as acidic as battery acid. There are even microbes capable of withstanding intense radiation, heavy metals, and complete darkness for centuries. In some cases, extremophiles use entirely different metabolic processes, generating energy from sulfur, iron, or methane instead of sunlight or organic matter. These super-resilient microbes have fascinated scientists, not just for their novelty, but for what they reveal about the potential for life on other planets. If life can exist in the boiling vents at the bottom of the sea or inside nuclear waste, could it also survive on Mars or Europa? Extremophiles expand our understanding of biology and challenge our definitions of what life truly is.

In addition to spreading across vast environments, microbes also organize themselves into complex communities known as biofilms. A biofilm is a slimy, protective matrix produced by microbes that allows them to stick to surfaces and each other. This formation is not random—biofilms represent cooperative behavior among microbes, enabling them to share nutrients, resist antibiotics, and survive in harsh conditions. Biofilms can be found in nature, such as on river rocks or the surface of leaves, as well as in human-made environments like water pipes, medical implants, and even toothbrushes. In the human body, biofilms are behind persistent infections, such as plaque on teeth or chronic wounds that resist healing. What makes biofilms so fascinating is their community-like behavior—microbes within them can communicate using chemical signals and adjust their behavior based on environmental cues. They behave more like a tiny society than isolated cells, showing a surprising level of organization and resilience.

Microbes do not just live on the fringes of our awareness—they are the invisible architects of life on Earth. From soil to sky, from boiling vents to our very skin, they shape the environments they inhabit and, in turn, shape us.

Understanding where microbes live is essential for everything from agriculture and medicine to space exploration and environmental conservation. For the youth of today, this knowledge fosters not only curiosity but also a profound respect for the hidden forces that keep our planet alive. The next time we look at a grain of dirt, a puddle, or even our hands, we might pause to wonder: what invisible kingdoms are thriving there? In the smallest of spaces and the harshest of conditions, life finds a way—and microbes lead the charge.

End of the Chapter

Chapter-10. Microbiomes – The Invisible World Inside us

Discover the vast hidden universe living inside you. This chapter explores how trillions of microscopic partners shape digestion, immunity, mood, and overall health – revealing that caring for yourself also means caring for the invisible ecosystem within.

Deep within our bodies lives an invisible ecosystem so vast and complex that it outnumbers our own cells by trillions. This hidden world, known as the human microbiome, is a community of bacteria, viruses, fungi, and other microorganisms that live on and inside us—particularly in our gut. Far from being mere passengers, these microbes play essential roles in digestion, immunity, mental health, and overall well-being. This chapter invites young readers to explore how the microbiome operates, how it helps us stay healthy—or sick—and how lifestyle choices like diet and hygiene influence its balance. As we journey inward, we discover that some of the most important partnerships in our lives are happening on a microscopic level, every single moment.

The human microbiome is the collective term for all the microorganisms that reside in and on the human body. These microbes inhabit various environments such as the skin, mouth, nose, and especially the digestive tract. While the gut microbiome has received the most scientific attention, each body site hosts a unique microbial community suited to its conditions—moist, dry, acidic, or oily. The number of microbes in a person's body is staggering—over 100 trillion, outnumbering human cells by a ratio of about 1.3 to 1. But what's truly fascinating is that the genetic material of these microbes—known as the microbiome genome—is far greater than our own, giving us access to countless additional biochemical tools. The microbiome helps us digest food, absorb nutrients, synthesize vitamins like B12 and K, and even protect us from invading pathogens by crowding them out. It's not just a background feature of our biology—it's an active, dynamic part of who we are.

One of the most intriguing discoveries in recent years is the gut-brain connection, a biological highway that links the intestinal microbiome with our central nervous system. Through chemical signaling, the gut microbiota influence the production of neurotransmitters such as serotonin and dopamine, which affect mood, sleep, and behavior. In fact, over 90% of serotonin, often called the "feel-good" hormone, is produced in the gut. Scientists have even begun referring to the gut as the "second brain" because of its profound impact on mental health. Conditions like anxiety, depression, and stress have all been linked to imbalances in the gut microbiome. Furthermore, the immune system—roughly 70% of which resides in the gut—is trained and regulated by the presence of beneficial microbes. They help the body distinguish between harmful invaders and harmless substances, reducing inflammation and autoimmune responses. This gut-immune-brain triangle shows just how interconnected our mental, physical, and microbial health really is. Not all microbes in the human body are friendly, and maintaining a balance between helpful and harmful species is critical. "Good" microbes—like Lactobacillus and Bifidobacterium—help digest food, strengthen the intestinal lining, and keep the immune system alert but not overreactive. These beneficial bacteria often live in harmony, sharing resources and keeping each other in check. However, when this balance is disrupted—by illness, antibiotics, stress, or poor diet—opportunistic "bad" microbes can multiply. Pathogens such as Clostridium difficile or Candida albicans may overgrow, leading to infections, inflammation, or chronic health problems. This state of imbalance is known as dysbiosis, and it has been associated with conditions ranging from obesity and diabetes to allergies and irritable bowel syndrome. Understanding the dynamic between good and bad microbes helps us see that health is not about eliminating all germs, but rather about cultivating a diverse and balanced microbial community.

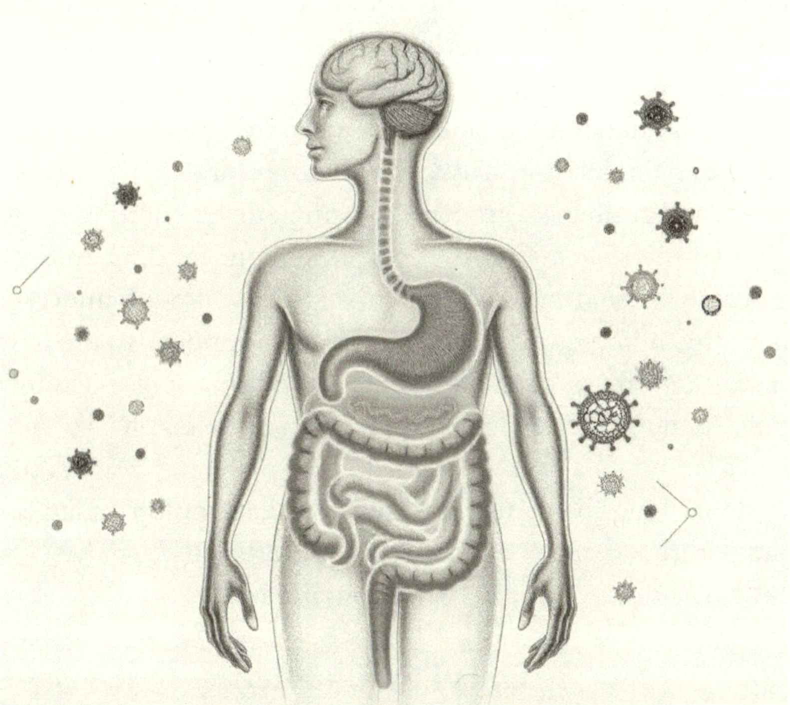

Diet plays a major role in shaping the microbiome, and this has led to growing interest in probiotics and microbiome-friendly nutrition. Probiotics are live beneficial microbes that, when consumed in adequate amounts, can improve gut health. These are found in fermented foods like yogurt, kefir, kimchi, sauerkraut, and tempeh. They help repopulate the gut with helpful bacteria, especially after taking antibiotics or recovering from illness. Prebiotics—non-digestible fibers found in foods like garlic, onions, bananas, and oats—feed these good microbes and support their growth. A diet rich in fiber, whole grains, fruits, and vegetables encourages microbial diversity, while heavily processed foods, artificial sweeteners, and excess sugar can reduce it. Regular physical activity, good sleep, and managing stress also influence microbiome balance. Together, these lifestyle factors contribute to what scientists call "microbial resilience"—the ability of your internal ecosystem to bounce back from disruptions. For young people, developing these habits early is like nurturing a lifelong partnership with an internal community that's always working behind the scenes.

The human microbiome is not just a feature of our biology—it is a dynamic, living network that contributes to our identity, health, and resilience. From influencing our emotions to training our immune system and breaking down our food, microbes are intricately woven into our daily lives. The more we understand about this microscopic world, the better equipped we are to care for it—and ourselves. For youth, exploring the microbiome opens a new dimension of self-awareness, emphasizing that health isn't only about what we do externally, but also about the life flourishing inside us. The invisible allies within us are proof that nature's most powerful tools are often the smallest—and that our bodies are not just hosts, but homes.

End of the Chapter

Chapter-11. Microbes and Diseases – How Infections Happen

Uncover the hidden battles between microbes and the human body. This chapter explains how infections begin, spread, and are defeated—empowering you with knowledge that turns everyday habits into powerful tools for protecting health and saving lives.

Microbes are not always our silent helpers. Sometimes, they turn into fierce invaders capable of causing illness and even death. From the common cold to life-threatening diseases like tuberculosis, malaria, and COVID-19, microorganisms are behind many of the health challenges humanity has faced across history. Understanding how these infections happen—how microbes invade, spread, and trigger our body's defense systems—is crucial for protecting ourselves and our communities. This chapter offers young readers a deep dive into the world of microbial diseases, revealing the many ways infections begin, how our immune system fights back, and how simple actions like handwashing and wearing masks can stop the spread. In a world more connected than ever, knowledge of microbial infections isn't just scientific curiosity—it's a tool for survival, safety, and smart citizenship.

Microbial diseases come in many forms, each caused by a different type of pathogen with its own unique strategy for infecting the human body. Bacterial diseases are caused by single-celled prokaryotic organisms. These infections range from mild ailments like strep throat to serious conditions like tuberculosis, pneumonia, and bacterial meningitis. Bacteria can multiply rapidly and often release toxins that damage tissues or interfere with normal bodily functions.

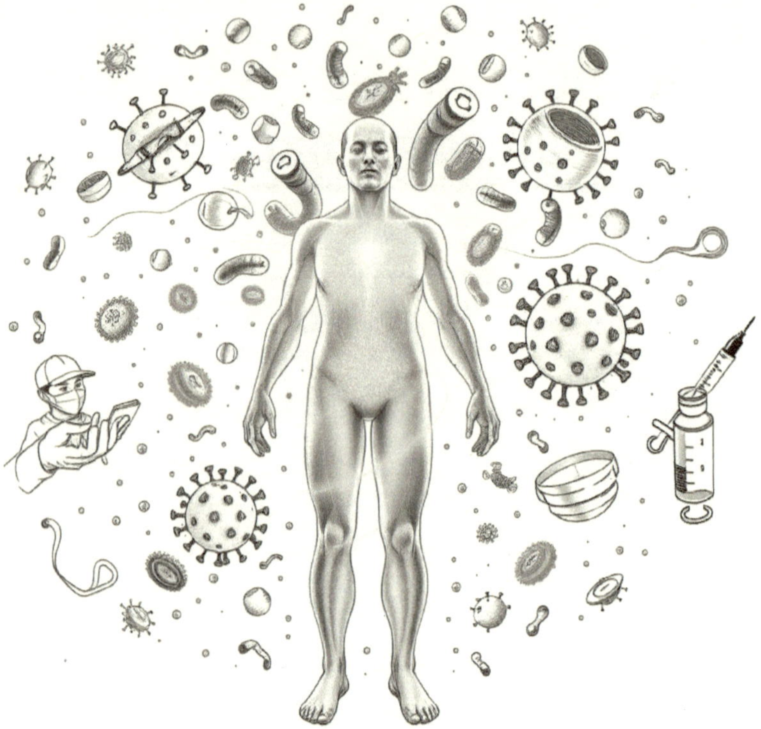

Viral diseases, on the other hand, are caused by tiny infectious particles that aren't fully alive. Viruses like the influenza virus, HIV, or SARS-CoV-2 (which causes COVID-19) invade host cells and hijack their machinery to reproduce. Because they hide inside our own cells, they can be difficult to detect and destroy. Fungal diseases, often seen as skin infections like athlete's foot or ringworm, can become severe in people with weakened immune systems, leading to conditions like systemic candidiasis or aspergillosis. Protozoan diseases, caused by single-celled eukaryotes, include malaria, amoebiasis, and sleeping sickness. These microbes often live in water or are spread through insects, and they use complex life cycles to evade the immune system and persist in the body.

Infectious microbes can spread in many different ways, and each mode of transmission brings its own challenges for prevention and control. Direct transmission involves close contact with an infected person—through touching, kissing, coughing, or sneezing. Diseases like the flu or the common cold often spread this way. Indirect transmission can happen via surfaces (fomites) that harbor microbes, such as door handles, phones, or shared utensils. Airborne transmission involves microscopic droplets or aerosols that linger in the air, allowing diseases like measles or COVID-19 to spread without direct contact. Some diseases, like cholera or hepatitis A, are spread through contaminated water or food, while others, like malaria and dengue, are transmitted by insect vectors such as mosquitoes. The symptoms of microbial infections vary depending on the pathogen but commonly include fever, fatigue, pain, inflammation, coughing, and digestive issues. These symptoms are often signs of the body's immune response trying to eliminate the invader. Understanding how diseases spread is the first step toward stopping them, empowering youth to become more conscious about hygiene, behavior, and the health of those around them.

When microbes enter the body, the immune system springs into action to defend against them. This defense system has two main arms: natural (or innate) immunity and acquired (or adaptive) immunity. Natural immunity is the body's first line of defense—it includes physical barriers like skin, mucous membranes, stomach acid, and cells like macrophages that quickly attack anything foreign. Though fast, this response is not highly specific. Acquired immunity, however, is much more targeted. It involves specialized cells called lymphocytes—B cells and T cells—that recognize specific pathogens. B cells produce antibodies that neutralize the invader, while T cells destroy infected cells. Once the body has fought off an infection, it often "remembers" the microbe, making future responses faster and more effective. This memory is what makes vaccines so powerful—they expose the immune system to a harmless version of the pathogen, training it to fight the real thing later. Without the immune system, even minor infections could become fatal. With it, we have one of the most sophisticated defense networks nature has ever created.

In the battle against infectious diseases, some of the most powerful tools are surprisingly simple. Hand hygiene is one of the most effective and accessible public health practices available. Washing hands with soap and water for at least 20 seconds removes not just visible dirt but also millions of microbes that could cause disease. Hand sanitizer with at least 60% alcohol is a good alternative when soap isn't available. During outbreaks or pandemics, wearing masks—especially in crowded or poorly ventilated spaces—can dramatically reduce the spread of airborne microbes. Masks act as barriers, trapping respiratory droplets that may contain viruses or bacteria. Public health strategies such as vaccination campaigns, quarantine measures, safe food handling, clean water access, and public awareness efforts also play a huge role in preventing disease transmission. Young people, by understanding and adopting these practices, can help protect not only themselves but also their families, schools, and communities. Public health is not just the responsibility of doctors or governments—it's a shared mission that begins with individual choices.

Infectious diseases are a natural part of the living world, but with knowledge, vigilance, and science-backed habits, we can control and even eliminate many of them. Microbes may be tiny, but their impact is monumental—sometimes destructive, often complex, and always fascinating. For today's youth, understanding how infections happen opens a door to smarter living and deeper empathy. It shows that health is interconnected—that one person's choices affect many others—and that even small habits, like washing hands or covering a cough, can be acts of care and protection. In a world where microbial threats may always exist, the greatest defense is not fear, but awareness.

End of the Chapter

Chapter-12.
Microbes in Food – From Kitchen Ferments to Factory Floors

Taste the science behind your favorite foods. This chapter reveals how microbes transform simple ingredients into flavorful, nutritious creations and how science keeps food safe – showing that every bite is a collaboration between humans and microscopic chefs.

Food is more than just fuel—it's culture, tradition, innovation, and science on a plate. And hidden behind many of the tastes and textures we love are the remarkable actions of microbes. These microscopic organisms have been helping humans prepare, preserve, and improve food for thousands of years, often without our awareness. From bubbling yogurt to crusty bread, tangy pickles to creamy cheese, microbes are the unseen chefs in our kitchens. But their role doesn't stop at home; in modern food industries, microbial science has become central to ensuring food safety, preventing spoilage, and enhancing nutrition. This chapter uncovers how fermentation works, how microbes produce enzymes that transform raw ingredients, and how microbiologists protect our food from harmful pathogens on a factory scale. For youth, it's a delicious discovery of how the smallest living things shape what we eat—and how we can use them wisely in the kitchens and industries of tomorrow.

Fermentation is one of the oldest and most valuable microbial processes used in food preparation. It's a natural chemical process where microbes like bacteria and yeast break down sugars and other molecules into simpler compounds—often producing gases, acids, or alcohol as byproducts. These changes not only preserve food but also enhance its flavor, texture, and digestibility. In yogurt production, for example, bacteria such as Lactobacillus bulgaricus and Streptococcus thermophilus consume lactose (milk sugar) and convert it into lactic acid, giving yogurt its tangy taste and thick texture. In bread-making, yeast (Saccharomyces cerevisiae) ferments the sugars in flour, producing carbon dioxide that causes the dough to rise and become airy. Cheese-making involves the fermentation of milk using specific bacteria and molds, which help in curdling, ripening, and flavor development. Pickles are made when vegetables are immersed in a salty or acidic solution, encouraging lactic acid bacteria to thrive and create the sour taste while preserving the produce for long periods. Fermentation is not just a culinary art—it's a partnership between humans and microbes that dates back to ancient civilizations.

Microbial metabolism lies at the heart of many transformations in food. When microbes consume sugars, starches, or proteins, their metabolic processes generate enzymes—biological catalysts that speed up chemical reactions. These enzymes break down complex molecules into simpler ones, altering the food's texture, flavor, and nutritional properties. For instance, the enzyme amylase breaks down starch into simple sugars, a process crucial in brewing and baking. Protease breaks down proteins, softening meat or helping cheese ripen. In soy sauce and miso, molds like Aspergillus oryzae produce a cocktail of enzymes that ferment soybeans into umami-rich condiments. These microbial actions not only produce flavorful results but also make certain foods easier to digest and their nutrients more bioavailable. Enzymes produced by microbes are now also widely used in industrial food processing—for clarifying fruit juices, tenderizing meat, and even reducing lactose in dairy products for lactose-intolerant consumers. Understanding microbial metabolism helps us innovate new food products while preserving traditional ones.

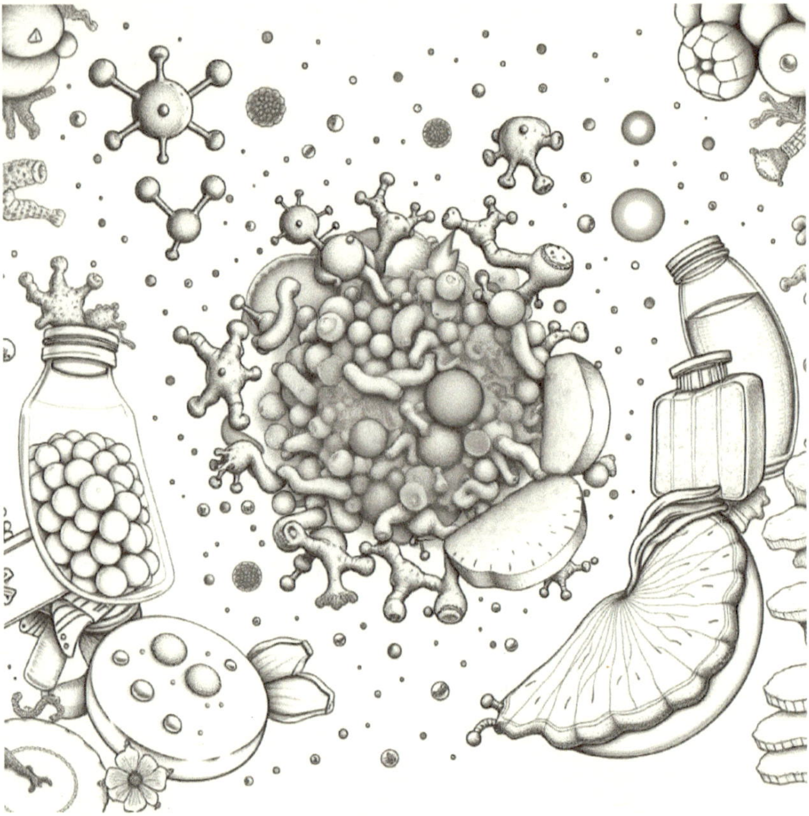

While many microbes are beneficial, others can cause foodborne illnesses if not properly controlled. This is where pasteurization plays a crucial role. Named after Louis Pasteur, who pioneered the method, pasteurization involves heating liquids like milk, juice, or wine to specific temperatures for a set time to kill harmful pathogens without significantly affecting the food's taste or nutritional value. Pasteurization doesn't make food sterile, but it greatly reduces the risk of diseases like tuberculosis, salmonellosis, and listeriosis that were once common in raw milk or underprocessed foods. In modern times, variations of this method—like ultra-high temperature (UHT) treatment and flash pasteurization—have extended the shelf life of beverages while maintaining safety. Beyond pasteurization, other food safety measures include refrigeration, freezing, vacuum sealing, and chemical preservatives—all designed to slow microbial growth and protect consumers. These interventions demonstrate the fine balance food microbiologists maintain: preserving beneficial microbes while eliminating dangerous ones.

In industrial settings, microbiology is applied on a massive scale to enhance food production, monitor quality, and prevent spoilage. Industrial food microbiologists work in labs and manufacturing plants to ensure that the microbial content of food stays within safe limits. They use microbial cultures to mass-produce fermented goods like cheese, vinegar, beer, and probiotic beverages. In quality control labs, samples are tested for contamination by pathogens like E. coli, Salmonella, and Listeria, which can enter the food supply chain through raw ingredients, equipment, or improper handling. Spoilage microbes—though often non-pathogenic—can cause food to look, smell, or taste bad, leading to waste and economic loss. To combat this, industries use techniques such as rapid microbial detection, packaging innovations, and natural preservatives like nisin and bacteriocins—compounds made by bacteria that inhibit spoilage organisms. Advances in biotechnology have also enabled the development of genetically modified microbes that improve shelf life, enhance nutrition, or reduce allergens. The food industry's relationship with microbes is a blend of vigilance and creativity, ensuring that what we eat is both safe and satisfying.

The world of food microbiology is a blend of tradition and innovation, ancient techniques and cutting-edge science. From the grandmother's pickle jar to the sterile halls of a dairy processing plant, microbes are at work—flavoring, preserving, and sometimes threatening our food. For young minds, this field offers not only scientific insight but also endless opportunity for invention. Learning how microbes interact with what we eat invites curiosity, respect for nature, and the possibility to build a future where food is healthier, tastier, and more sustainable. Microbes may be invisible, but in the world of food, their impact is felt with every delicious bite.

End of the Chapter

Chapter-13.
Microbes in Agriculture : Nature's Green Allies

Dig into the hidden partnership beneath every thriving crop. This chapter reveals how microbes enrich soil, protect plants, and power sustainable farming – showing how nature's tiniest allies help feed the world and shape the future of agriculture.

Beneath our feet lies an invisible world bustling with life – one that powers the food we eat, the plants we grow, and the health of the planet itself. This hidden network of microbes plays an extraordinary role in supporting agriculture, yet most of us are unaware of their contributions. These microscopic organisms are not pests or problems – they're powerful allies. From enriching soil with vital nutrients to defending crops against pests and diseases, microbes form the foundation of sustainablè, eco-friendly farming. This chapter takes young readers on a journey into the soil and beyond to discover how bacteria, fungi, and other tiny life forms help feed the world naturally. As the next generation faces the dual challenges of food security and environmental preservation, learning to work with microbes – not against them – could unlock smarter, greener agricultural solutions for the future.

One of the most remarkable contributions microbes make to agriculture is through biofertilizers – natural, living organisms that enrich the soil with nutrients and improve plant growth. Among the most well-known biofertilizers are nitrogen-fixing bacteria such as Rhizobium. These bacteria form symbiotic relationships with leguminous plants like peas, beans, and lentils, colonizing nodules on their roots. Inside these nodules, Rhizobium converts atmospheric nitrogen (which plants cannot use) into ammonia, a form of nitrogen that plants can absorb and use for growth. This process not only nourishes the host plant but also boosts soil fertility for future crops. Other bacteria like Azospirillum and Azotobacter, and phosphate-solubilizing microbes like Pseudomonas, further enhance nutrient availability in the soil. Unlike chemical fertilizers that can harm ecosystems with overuse, biofertilizers work in harmony with nature, building soil health, improving water retention, and reducing environmental pollution. For young farmers and scientists, these microbial partners offer a natural alternative to synthetic inputs—restoring soil vitality and long-term sustainability.

Microbes also play a crucial role in protecting crops from pests and diseases through the use of biopesticides—natural agents derived from bacteria, fungi, viruses, or other microbes that target harmful insects and pathogens. A famous example is Bacillus thuringiensis (Bt), a soil bacterium that produces proteins toxic to certain insect larvae but harmless to humans, animals, and beneficial insects. When used as a spray or seed coating, Bt reduces the need for chemical pesticides and lowers the risk of environmental damage. Other microbes like Trichoderma fungi and Beauveria bassiana act as natural enemies of plant pathogens or insect pests by outcompeting them or releasing antifungal or insecticidal compounds. Unlike broad-spectrum chemical pesticides, which often kill indiscriminately, microbial biopesticides are targeted, biodegradable, and environmentally safe. They allow farmers to adopt integrated pest management practices that minimize chemical use while preserving biodiversity. For youth interested in agriculture or ecology, biopesticides showcase how smart, biology-based approaches can protect crops without poisoning the planet.

Another area where microbes show their strength is composting—the transformation of organic waste into rich, nutrient-dense fertilizer through microbial action. Composting relies on decomposer microbes, including bacteria and fungi, to break down plant residues, kitchen scraps, animal manure, and other biodegradable materials into humus. These microbes digest complex organic matter into simpler compounds, generating heat in the process and killing off weed seeds and pathogens. The final product is dark, crumbly, and odorless—perfect for enriching garden beds, potted plants, and farm fields. Microbial composting not only recycles waste that would otherwise end up in landfills, but also restores soil texture, boosts microbial biodiversity, and reduces reliance on chemical fertilizers. New innovations in microbial composting include the use of specialized microbial inoculants—preparations containing selected decomposer strains that speed up the composting process and improve quality. By composting with microbes, even young gardeners can play a hands-on role in closing the nutrient loop and turning trash into treasure.

Microbial inoculants—also known as microbial consortia or bioinoculants—are preparations containing beneficial bacteria, fungi, or a mix of both, used to boost plant health and crop productivity. These products are applied to seeds, roots, or soil to introduce helpful microbes that enhance nutrient uptake, stimulate growth, and suppress disease. One promising group of inoculants are mycorrhizal fungi, which form symbiotic relationships with plant roots, extending their reach deep into the soil and improving access to water and minerals. Similarly, Plant Growth Promoting Rhizobacteria (PGPR) such as Pseudomonas fluorescens and Bacillus subtilis release plant hormones, mobilize nutrients, and produce antibiotics that inhibit harmful pathogens. In drought-prone regions, certain inoculants help crops tolerate water stress by improving root architecture and water use efficiency. With climate change challenging traditional farming methods, microbial inoculants offer a cutting-edge, eco-friendly solution to boost yield while preserving soil and ecosystem health. For future agripreneurs and environmentalists, they represent the frontier of bio-based agriculture – where success grows from the soil up.

Microbes in agriculture are more than microscopic helpers – they are nature's architects, quietly engineering healthier soils, safer crops, and more sustainable food systems. As the global population rises and environmental pressures intensify, the need for clean, efficient, and regenerative farming practices has never been greater. Microbes offer us ancient wisdom in modern form, showing that the answers to many of our agricultural problems already exist—beneath our feet and within the roots. For youth exploring science, farming, or environmental stewardship, the message is clear: by teaming up with nature's tiniest allies, we can cultivate a greener, more resilient world—one microbe at a time.

End of the Chapter

Chapter-14. Microbes for a Cleaner Planet – Environmental Solutions

Discover how tiny organisms become mighty environmental heroes. This chapter reveals how microbes clean pollution, recycle waste, purify water, and fight climate change – showing that some of Earth's most powerful solutions come from its smallest life forms.

In a world increasingly overwhelmed by pollution, waste, and climate change, the smallest organisms on Earth are emerging as some of our most powerful allies. Microbes – once seen merely as agents of disease – are now recognized for their extraordinary potential to heal the planet. These invisible workers play a key role in cleaning the environment, recycling nutrients, breaking down pollutants, purifying water, and even influencing the Earth's climate. This chapter takes young readers deep into the environmental roles of microbes, exploring how their ancient survival strategies are now being harnessed to create modern solutions. As the next generation of environmental guardians, youth can be inspired by the microbial world's capacity to restore balance, showing that saving the planet may not always require heavy machinery or grand inventions – sometimes, it starts with the tiniest forms of life.

Microorganisms are the world's natural recyclers, and decomposition is one of their most fundamental and impactful roles. Every leaf that falls, every dead animal, every piece of organic waste is broken down by microbial decomposers –

bacteria, fungi, and actinomycetes that live in the soil and air. These microbes feed on organic matter, breaking it into simpler compounds like carbon dioxide, water, and nutrients such as nitrogen, phosphorus, and potassium.

These elements are then returned to the ecosystem, becoming available to plants and other organisms, completing nature's nutrient cycles. Without decomposition, the Earth would be buried under layers of waste, and life-giving nutrients would be locked away.

In forests, compost heaps, wetlands, and even deserts, microbes quietly transform death into life, ensuring that ecosystems remain fertile and balanced. For young minds, understanding microbial decomposition is a lesson in the interconnectedness of life—how nothing in nature is wasted, and how even decay is a form of renewal.

Beyond natural recycling, microbes are now used in bioremediation—the process of using living organisms to clean up environmental messes caused by human activity. Oil spills, chemical leaks, pesticide residues, and industrial pollutants can devastate ecosystems, but certain microbes have the unique ability to digest these harmful substances.

In marine oil spills, for example, hydrocarbon-degrading bacteria such as Alcanivorax and Pseudomonas feed on petroleum, breaking it down into less toxic compounds. These microbes are often applied directly to contaminated water, helping restore marine life and water quality. In soil polluted with heavy metals or toxic chemicals, bioaugmentation—adding specialized microbes—can help detoxify and restore fertility.

Even radioactive waste sites have been found to host resilient extremophile microbes capable of immobilizing radioactive elements. Bioremediation is a powerful, low-cost, and eco-friendly alternative to chemical clean-ups, and it reflects nature's own method of healing. For youth concerned with pollution and conservation, this field offers hope and practical pathways to environmental repair using biology, not bulldozers.

Microbes also play a central role in managing urban waste and purifying water—two critical challenges in a growing world. In sewage treatment plants, microbial communities break down organic waste in multiple stages.

In the primary treatment stage, solid waste is removed, while the secondary treatment uses aerobic bacteria to consume organic matter in wastewater.

These bacteria thrive in oxygen-rich tanks, turning harmful compounds into harmless byproducts like carbon dioxide and sludge, which can be used as fertilizer.

In the final stage, disinfection using UV light or chlorine ensures that remaining pathogens are eliminated before water is released into rivers or reused. In septic tanks and decentralized treatment systems, anaerobic microbes (those that live without oxygen) play a similar role, breaking down waste in low-oxygen conditions. Additionally, microbes are essential in biofilters that clean water in natural wetlands or artificial ponds, creating sustainable solutions for rural areas. Teaching youth how microbes purify water instills a sense of awe and responsibility—recognizing that clean water is not magic, but microbial science at work.

Microbes also have a surprising connection to the global climate. Through their roles in the carbon and nitrogen cycles, microbes influence the levels of greenhouse gases in the atmosphere. Soil microbes help regulate carbon storage by decomposing plant matter and releasing carbon dioxide—a natural part of the cycle.

However, some microbes also fix carbon, storing it in stable forms in the soil. Others, like methanotrophic bacteria, consume methane, a potent greenhouse gas. In wetlands and rice paddies, where methane-producing microbes are active, managing microbial communities can significantly reduce emissions. Similarly, nitrogen-fixing microbes and denitrifiers affect the levels of nitrous oxide, another greenhouse gas.

Climate scientists and ecologists are now exploring ways to manage microbial populations to enhance carbon capture, reduce emissions, and build climate-resilient soils. For youth interested in climate action, understanding microbial roles in Earth's systems opens a window into one of the most powerful, natural tools for environmental sustainability.

Microbes are not just ancient survivors—they are environmental innovators. They clean our messes, recycle our waste, filter our water, and regulate our climate, often without us even noticing.

Their incredible versatility, resilience, and adaptability make them indispensable allies in building a greener, cleaner future. By exploring their hidden powers, young readers can begin to see the Earth not as something broken, but as something constantly being repaired—quietly, persistently, and powerfully—by the smallest life forms on the planet. In the era of environmental crises, microbes offer more than inspiration—they offer real-world solutions waiting to be harnessed by the minds and hands of a new generation.

End of the Chapter

Chapter-15.
Medical Microbiology – Healing with the Smallest Healers

Enter the world where microbes become medicine. This chapter uncovers how tiny organisms inspire antibiotics, vaccines, and revolutionary therapies – revealing how the smallest healers are shaping the future of human health.

Throughout history, microbes were seen as villains—causing disease, epidemics, and fear. But the tide turned when science began to reveal that these very same microscopic beings could also heal. From antibiotics to vaccines, modern medicine owes some of its most powerful tools to the microbial world. Medical microbiology, the branch of science that studies microbes to understand and fight illness, is one of the most exciting and life-saving fields in all of biology. This chapter explores how discoveries like penicillin changed history, how antibiotic resistance is creating new challenges, and how today's innovations—like gene therapy, phage treatment, and microbiome medicine—are shaping the future of health. For young learners, it's a journey into the unseen battles between microbes and the human body—and how we can harness the smallest healers to defend and repair our health in the most extraordinary ways.

The discovery of antibiotics stands as one of the greatest breakthroughs in the history of medicine, and it began with a mold. In 1928, Alexander Fleming accidentally discovered that a mold called Penicillium notatum could kill bacteria. He observed that the area around the mold on a petri dish was clear of Staphylococcus bacteria.

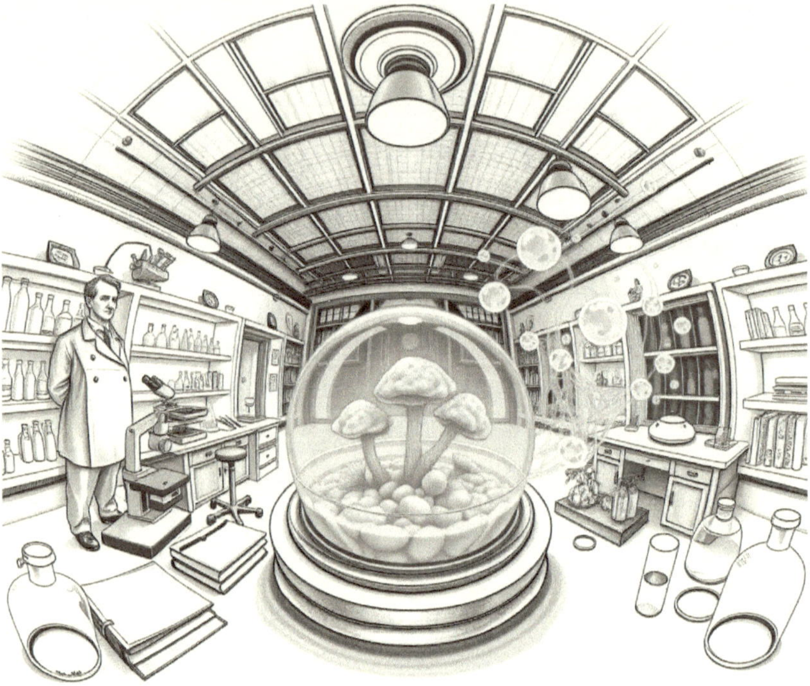

This led to the development of penicillin, the world's first true antibiotic—a substance that could destroy or inhibit the growth of harmful bacteria inside the human body.

By the 1940s, penicillin was saving countless lives on the battlefields of World War II, and it sparked a golden age of antibiotic discovery.

Soon, other antibiotics like streptomycin, tetracycline, and erythromycin followed, revolutionizing the treatment of bacterial infections such as pneumonia, tuberculosis, and syphilis. These life-saving drugs, many of which are still derived from soil microbes, marked the beginning of modern medical microbiology. They taught us that microbes could be both enemies and allies—causing illness but also curing it. However, the miracle of antibiotics is now under serious threat due to a growing crisis: antimicrobial resistance (AMR). This occurs when bacteria evolve to resist the effects of antibiotics, making treatments less effective and infections harder to cure. Overuse and misuse of antibiotics—in humans, animals, and agriculture—have accelerated this process. As a result, "superbugs" like Methicillin-resistant Staphylococcus aureus (MRSA) and drug-resistant tuberculosis have emerged, causing infections that are increasingly difficult or even impossible to treat. These resistant strains can spread quickly, especially in hospitals and crowded environments. AMR threatens to undo decades of medical progress, potentially making routine surgeries, childbirth, and minor injuries deadly once again. Scientists are now racing to find new antibiotics, develop rapid diagnostics to guide their use, and promote global awareness about responsible antibiotic practices. For today's youth, understanding AMR is essential—not just as future doctors or scientists, but as informed citizens who play a role in preserving the effectiveness of life-saving medicines.

Microbes have also been instrumental in the development of vaccines, diagnostics, and even revolutionary therapies like gene editing. Vaccines are one of the most powerful public health tools ever created, and many rely on weakened or inactivated microbes to train the immune system. The smallpox vaccine, the first of its kind, used cowpox virus to protect against a deadly disease that once ravaged humanity. Today, vaccines against polio, hepatitis, measles, and COVID-19 protect billions of people worldwide—and many of these vaccines are built using microbial technology, including viruses engineered to deliver genetic material. Diagnostic tools also depend on microbes. Techniques like PCR (polymerase chain reaction), which allow doctors to detect infections by amplifying microbial DNA, have become central to identifying viruses and bacteria rapidly and accurately. In gene therapy, modified viruses are used as delivery vehicles to insert corrected genes into human cells, offering hope for treating genetic disorders like sickle cell anemia and certain cancers. These breakthroughs reflect the vast potential of microbes not just as causes of disease, but as tools for its cure.

Another groundbreaking frontier in medical microbiology is microbiome medicine—the idea that the trillions of microbes living inside our bodies play a key role in our overall health. Rather than simply eliminating bacteria, doctors are now studying how to support and restore beneficial microbes in the gut, skin, lungs, and even the brain. Disruptions in the microbiome have been linked to conditions like obesity, asthma, depression, and autoimmune diseases. New treatments aim to reintroduce "good" microbes through probiotics, prebiotics, and even fecal microbiota transplants (FMT), where healthy gut bacteria from a donor are transferred to a patient. At the same time, phage therapy—an old idea gaining new attention—is being revisited as an answer to antibiotic resistance. Bacteriophages, or phages, are viruses that infect and kill specific bacteria without harming human cells. Unlike antibiotics, which can disrupt entire microbial communities, phages target their bacterial hosts with incredible precision. Scientists are now developing personalized phage treatments for patients with resistant infections, and early results are promising. These new directions in microbiome and phage therapy signal a shift from fighting microbes to collaborating with them—a concept that redefines medicine in the 21st century.

Medical microbiology reveals a powerful truth: within the smallest life forms lie the greatest healing possibilities. Microbes are not just threats to be feared—they are partners in the science of health, capable of inspiring some of the most innovative medical tools of our time. From antibiotics and vaccines to gene therapy and phages, the solutions to many of our most pressing health challenges may come not from outside forces, but from deep within the microbial world. For young readers, this chapter opens the door to a future where healing doesn't always come in the form of a pill, but perhaps in a drop of bacteria, a breath of a virus, or a precisely tuned microbial cocktail. Embracing this science means stepping into a future where biology, technology, and imagination merge—to heal, protect, and transform lives.

End of the Chapter

Chapter-16.
Microbes in Biotechnology Engineering the Invisible

Step into the future where microbes become master engineers. This chapter reveals how tiny organisms are redesigned to create clean energy, sustainable materials, and advanced medical tools – showing how engineering the invisible is shaping tomorrow's world.

As we look to the future, one thing becomes clear: the smallest forms of life may hold the keys to solving some of humanity's biggest challenges. Microbes, often overlooked as mere germs or simple organisms, are being transformed into powerful tools by the hands of scientists and engineers. Through genetic manipulation, green energy technologies, sustainable materials, and even medical breakthroughs, microbes are becoming the ultimate bioengineers of the 21st century. This chapter dives deep into some of the most cutting-edge applications of microbiology—from CRISPR gene editing to microbial fuel cells and nanomedicine. Young readers will discover how these microscopic innovators are helping us build a healthier, cleaner, and more advanced world. For those seeking careers in science, sustainability, or health, microbial innovation offers a thrilling frontier where biology meets imagination, and nature meets next-gen technology.

One of the most revolutionary scientific breakthroughs in recent history is the development of genetic engineering—and microbes have played a central role in it. Genetic engineering allows scientists to alter the DNA of organisms to give them new abilities, such as producing medicine, resisting disease, or degrading pollutants. The crown jewel of this technology is CRISPR-Cas9, a gene-editing tool derived from a microbial defense system. Originally found in bacteria as a way to fend off invading viruses, CRISPR has been repurposed by researchers to precisely cut and modify genes in almost any living organism, including humans. This technology is already being used to study genetic diseases, develop treatments, and even attempt to eliminate inherited conditions. In microbes, CRISPR is being used to engineer bacteria that can produce rare antibiotics, secrete biofuels, or act as biosensors for environmental toxins. The beauty of CRISPR lies in its accuracy, simplicity, and affordability—making it accessible to labs across the world and opening doors for youth to engage with gene editing in classrooms, competitions, and future careers. What once seemed like science fiction is now a tool in a high school biotech lab, thanks to the brilliance hidden in microbial genomes.

In the search for sustainable energy, microbial fuel cells have emerged as a fascinating solution. These devices use specially selected bacteria to convert chemical energy from organic matter directly into electricity. As these microbes consume nutrients or waste materials, they produce electrons, which can be captured and used to generate power. The idea is both simple and revolutionary: instead of burning fossil fuels, let microbes digest waste and light our homes. In sewage treatment plants, microbial fuel cells are already being tested to power sensors and reduce energy costs. They are also being explored for remote areas where clean energy is scarce. One of the most exciting prospects is using microbial fuel cells in space missions, where waste can be turned into electricity and water in closed-loop life support systems. For youth passionate about clean energy and sustainability, this technology highlights how the smallest organisms can be harnessed to tackle our biggest energy problems—silently, efficiently, and renewably.

Plastic pollution is one of the most pressing environmental challenges of our time, and once again, microbes may offer the solution. Through genetic engineering and bioprocessing, scientists are using bacteria and algae to create bioplastics—eco-friendly materials that can replace conventional petroleum-based plastics. These bioplastics are biodegradable, renewable, and can be designed to break down harmlessly in the environment. One example is polyhydroxyalkanoates (PHAs), a family of biodegradable plastics produced naturally by certain bacteria as energy reserves. These can be molded into packaging, utensils, and containers—reducing our dependence on fossil fuels and minimizing plastic waste. Alongside this, microbes also contribute to the enzyme industry, where microbial enzymes are used in everything from detergents and textiles to paper and pharmaceuticals. Enzymes produced by genetically modified microbes can operate under extreme conditions, speeding up industrial processes while reducing chemical pollution. These innovations demonstrate that by partnering with microbes, we can build industries that are cleaner, greener, and smarter. For students interested in biotechnology, this field offers a powerful way to contribute to sustainability while working on exciting, real-world innovations.

Perhaps one of the most inspiring and futuristic applications of microbiology lies in the field of cancer research and nanomedicine. Scientists are now using engineered microbes to detect and even destroy cancer cells inside the human body. Some strains of bacteria can naturally home in on tumors due to the unique chemical environment they offer. By genetically modifying these bacteria, researchers are turning them into precision tools that deliver cancer-fighting drugs directly to tumors while sparing healthy cells. Meanwhile, nanomedicine—a field that works at the molecular and atomic scale—is using microbial products to develop nanoscale delivery systems. For example, nanoparticles made from bacterial proteins can carry drugs, monitor body chemistry, and trigger responses only when needed, reducing side effects and increasing effectiveness. In clinical trials, these technologies are already showing promise for treating hard-to-reach cancers and chronic illnesses. These advances symbolize a profound shift in medicine—from blunt-force approaches to targeted, personalized therapies made possible by microbes. For youth fascinated by health, innovation, and science fiction-like possibilities, this frontier offers limitless potential for discovery and impact.

Microbes are no longer confined to the lab or the soil—they are shaping the future of science, energy, health, and sustainability. Through CRISPR, fuel cells, bioplastics, and biomedical breakthroughs, these tiny organisms are proving that even the smallest life forms can create the biggest change. For the youth of today, understanding and engaging with microbial innovations opens the door to solving real-world problems and building a smarter, cleaner planet. The age of microbial technology is here—and it belongs to those curious enough to explore it.

End of the Chapter

Chapter-17.
Space Microbiology – Life Beyond Earth

Look beyond Earth to discover how tiny life forms may lead humanity into space. This chapter explores how microbes survive cosmic extremes and guide the search for extraterrestrial life – revealing that the path to the stars may begin with the smallest explorers.

As humanity reaches beyond our home planet into the vast expanse of space, one question continues to fascinate scientists and explorers alike: could life exist elsewhere in the universe? Surprisingly, the answers might lie not in grand alien civilizations but in the smallest forms of life—microbes. Microorganisms are incredibly resilient, able to survive extreme temperatures, radiation, and even the vacuum of space. Their survival capabilities are now central to a field called space microbiology, which studies how microbes behave in space environments and what their presence could mean for future space travel and the possibility of extraterrestrial life. This chapter explores the incredible ways microbes are shaping space exploration—from experiments on the International Space Station to theories of life on Mars. For young dreamers looking toward the stars, the journey of microbes in space proves that science fiction is rapidly becoming scientific fact.

Microbes have accompanied humans on nearly every mission into space—not as passengers, but as critical subjects of study. Onboard the International Space Station (ISS), astronauts work in a closed-loop environment where microbes play vital roles, both positive and potentially harmful. Some microbes help in recycling air, decomposing waste, or even assisting in plant growth experiments for future space farming. Others, however, pose risks by becoming more virulent or resistant to antibiotics in microgravity. This makes space microbiology essential for astronaut health and long-duration missions. For example, NASA has studied how bacteria like E. coli and Staphylococcus aureus change in space—growing faster, developing thicker cell walls, and sometimes producing more toxins. These findings are not only important for keeping astronauts safe but also help scientists better understand how life adapts to extreme conditions. Moreover, genetically engineered microbes are now being tested in space to produce materials like vitamins, medicines, and even bioplastics, making them miniature factories that could support human life on distant planets. For youth dreaming of careers in space science or medicine, the study of microbes in orbit is a launchpad into the future of interplanetary living.

One of the most mind-bending theories in space biology is panspermia—the idea that life on Earth might have originated from microbes that traveled through space, perhaps hitching a ride on asteroids or comets. This theory suggests that life did not begin on Earth but was seeded here from elsewhere in the cosmos, carried by hardy microorganisms capable of surviving interplanetary travel. While still debated, this concept has gained scientific interest because of real experiments showing that certain microbes can endure the harshest space conditions. For instance, Deinococcus radiodurans, often called the "superbug," can survive high levels of radiation, extreme cold, and dehydration. In controlled studies, microbes have been placed on the exterior of spacecraft, exposed to space for months, and returned alive. This opens the possibility that microbial life could survive a journey between planets—or even across solar systems—protected inside rocks or icy comets. The study of such microbes helps astrobiologists understand how life could spread in the universe, and how planetary protection protocols are needed to prevent Earth microbes from contaminating other worlds. For young minds, the idea that life is a traveling force throughout the galaxy turns the study of microbes into a truly cosmic adventure.

Astrobiology, the science of life beyond Earth, is now one of the most exciting frontiers in microbiology. As of now, we have no definitive evidence of alien life, but the most likely candidates for extraterrestrial organisms are not green humanoids—they're microbes. Scientists are focusing on environments that could support microbial life: subsurface oceans on Europa and Enceladus, underground water on Mars, and even the atmosphere of Venus. These places are inhospitable to humans but may harbor extremophiles—organisms that thrive in extreme conditions similar to those found in volcanic vents, acidic lakes, or frozen tundras on Earth. Robotic missions like NASA's Perseverance rover are equipped to search for microbial biosignatures—chemical or structural signs of past or present microbial life. If such evidence is found, it would fundamentally change our understanding of biology and our place in the universe. It would mean life is not a rare Earth phenomenon but a cosmic norm. For youth fascinated by the possibility of aliens, the tools of microbiology offer a real path to answering that question scientifically. The future of astrobiology may be led not by astronomers alone, but by microbiologists peering into the soil of distant planets.

Space microbiology transforms our view of life and survival. It teaches us that microbes are not just terrestrial beings but universal travelers, capable of adapting, thriving, and even pioneering life in space. Whether they're supporting astronauts on the ISS, sparking theories of life's origin, or guiding our search for aliens, microbes are proving essential to our interstellar story. For the next generation of explorers, scientists, and engineers, this field offers a thrilling invitation: study the smallest forms of life to unlock the biggest cosmic mysteries. The stars may be far, but microbes are already showing us how to reach them.

End of the Chapter

Chapter-18. Micro-Lab at School – DIY Experiments for Young Scientists

Turn curiosity into discovery with hands-on exploration. This chapter invites you to build your own micro-lab, use real scientific tools, and perform exciting experiments—proving that great scientific journeys can begin right at school.

Every great scientist starts with a simple experiment, a curious mind, and a space to explore. Microbiology may sound like a field limited to high-tech laboratories, but with the right approach, even school students can unlock its secrets through hands-on learning. This chapter is designed to turn classrooms, labs, and even kitchen corners into thriving centers of discovery. It introduces essential lab safety practices, teaches how to handle a microscope, and guides young readers through exciting, safe, and meaningful experiments they can perform themselves. These projects don't just demonstrate microbiological concepts—they cultivate observation, patience, creativity, and scientific thinking. For youth aspiring to become tomorrow's biologists, doctors, or environmentalists, these accessible activities are the first step into a much larger world of exploration and innovation.

The foundation of any experiment begins with safety. Whether in a classroom lab or home setup, students must understand the importance of wearing protective gear like gloves and goggles, avoiding the ingestion or inhalation of unknown substances, and always cleaning their workspace thoroughly before and after experiments. Microbes, though fascinating, can sometimes be harmful if grown or handled carelessly, so hygiene—such as not touching one's face during experiments and washing hands afterward—is essential. Clear labeling of petri dishes, proper disposal of biological materials (like spoiled bread or old cultures), and adult supervision during certain steps ensure that learning stays safe. Respecting microbes as both powerful allies and potential risks helps foster responsibility and awareness from an early stage, turning curious students into conscientious scientists.

A microscope is the magic key that unlocks the hidden world of microorganisms. Learning to use one begins with understanding its basic parts: the eyepiece (ocular lens), objective lenses, stage, light source, and focusing knobs. Students must handle slides gently and use clean lenses for accurate viewing. One of the simplest and most exciting tasks is preparing a wet mount—a glass slide where a small drop of water containing the sample is placed, covered with a cover slip, and viewed under the microscope. Onion skin cells, cheek cells, pond water, or yogurt all offer intriguing microscopic views. With patience and steady focus, students can see cells with visible nuclei, moving protozoa, or the rod and spiral shapes of bacteria. The first time a young scientist sees a living cell wriggle or divide under the lens, the microscopic world transforms from an abstract concept to a real, captivating universe.

One of the most classic and easy experiments is growing mold on bread. It's a visual, engaging way to observe fungal growth over time. To conduct this experiment, students take slices of bread—some moistened, some dry—and store them in sealed plastic bags placed in different conditions: a dark cabinet, a sunny window, the refrigerator. Over days, students will observe the emergence of fuzzy, colored mold colonies. They can record data daily—color, shape, growth rate—and compare how temperature and moisture affect microbial growth. This experiment teaches the basics of fungal biology, environmental influence on growth, and even introduces the concept of spore formation. Plus, it helps students understand real-world implications like food spoilage, hygiene, and preservation techniques.

Another fun and illuminating experiment is yeast and sugar fermentation. In this setup, students mix dry baker's yeast with warm water and sugar in a bottle, then place a balloon over the mouth of the bottle. Within minutes, the balloon begins to inflate as the yeast consumes the sugar and releases carbon dioxide gas—a byproduct of fermentation. This experiment demonstrates not only microbial metabolism but also the principles behind bread rising, alcohol production, and even renewable energy (bioethanol). It's a vivid, hands-on way to show how microbes "eat" and "breathe," turning simple ingredients into gas and energy. For youth, this brings abstract chemical reactions to life and underscores the power of tiny organisms to create visible, measurable change.

A powerful demonstration of hygiene and bacterial presence can be done through the handwashing and bacterial growth experiment. Here, students press their unwashed fingers onto nutrient agar plates, then repeat the process after washing hands with water, soap, or sanitizer. After leaving the plates in a warm place for a few days, bacterial colonies begin to grow—showing clear differences between cleaned and uncleaned samples. This experiment not only introduces the concept of culturing microbes but also reinforces the importance of hand hygiene in disease prevention. Students can compare results between different cleaning agents and techniques, encouraging critical thinking about health practices. In a world increasingly aware of infectious disease, this simple test drives home the message that daily habits have real biological consequences.

The Micro-Lab chapter offers more than just experiments—it offers empowerment. When students realize they can grow microbes, analyze results, and draw scientific conclusions, they take the first steps into authentic research. These activities are not only educational but also inspiring, allowing youth to feel the thrill of discovery and the responsibility of observation. Whether a future doctor, engineer, or ecologist, every path in science begins with a curious question—and a willingness to look through the microscope to find the answer.

End of the Chapter

Chapter-19. Becoming a Microbe Detective – Projects and Inquiry

Become a scientific detective in the invisible world. This chapter sparks curiosity with projects, models, and investigations that turn questions into discoveries – showing how young minds can explore microbes and think like real scientists.

Every great scientist begins with a question, a clue, and a curiosity to dig deeper. In the world of microbiology, being a "microbe detective" means developing the mindset of a researcher—observing patterns, asking why, and discovering how invisible organisms shape the world around us. This chapter is a hands-on invitation for young learners to go beyond textbooks and explore the microbial world through projects, models, experiments, and creative storytelling. Whether it's for a school science fair, an independent investigation, or a fun club activity, these project ideas spark inquiry, build problem-solving skills, and make science personal and real. With microbiology connecting to health, food, climate, and technology, students can find projects that match their interests while learning how to observe, experiment, and explain like true scientists.

Science fairs offer a platform where curiosity meets creativity. Students can take real-world questions and turn them into testable projects. A simple yet powerful idea is comparing hand sanitizer effectiveness. Students can test various sanitizers on bacterial growth using agar plates by applying finger imprints before and after sanitization. Another exciting project could be analyzing which household surface harbors the most bacteria—doorknobs, mobile phones, kitchen sponges, or school desks. Some students may choose to explore probiotic activity in yogurt brands by testing which one contains the most active cultures under a microscope.

Others might study the mold resistance of different bread types or investigate how sugar concentration affects fermentation in yeast. The goal isn't just to win a prize—but to learn how to formulate a hypothesis, conduct an experiment, collect data, and present findings. Science fairs give students the stage to not just display science but to do science—authentically and confidently.

Building 3D models of microbes brings biology into art and engineering. Students can construct large-scale models of bacteria, viruses, or fungi using common materials like clay, cardboard, paper-mâché, foam balls, and string.

A virus model, for instance, can showcase the protein capsid, genetic material, and tail fibers using colorful craft materials, helping others visualize its structure and function. A bacterial cell model can include a labeled nucleus (if applicable), cell wall, flagella, and plasmids.

These models are not only visually engaging but also deeply educational—they help learners internalize the forms, functions, and diversity of microbes.

Students can even explore creative formats like edible microbe models using candy or recycled material builds that highlight sustainability. Displaying such models in classrooms or science exhibitions allows students to teach others and take ownership of their learning, turning abstract textbook illustrations into memorable, tangible understanding.

Microbial tracking projects are powerful tools to study hygiene, behavior, and public health. Students can design experiments to trace how germs spread in their environment—by coating a student's hand with harmless, visible powder (like fluorescent lotion or glitter), then observing how it transfers onto doorknobs, keyboards, or books throughout the day. UV light can help make invisible particles glow, simulating real microbial transmission.

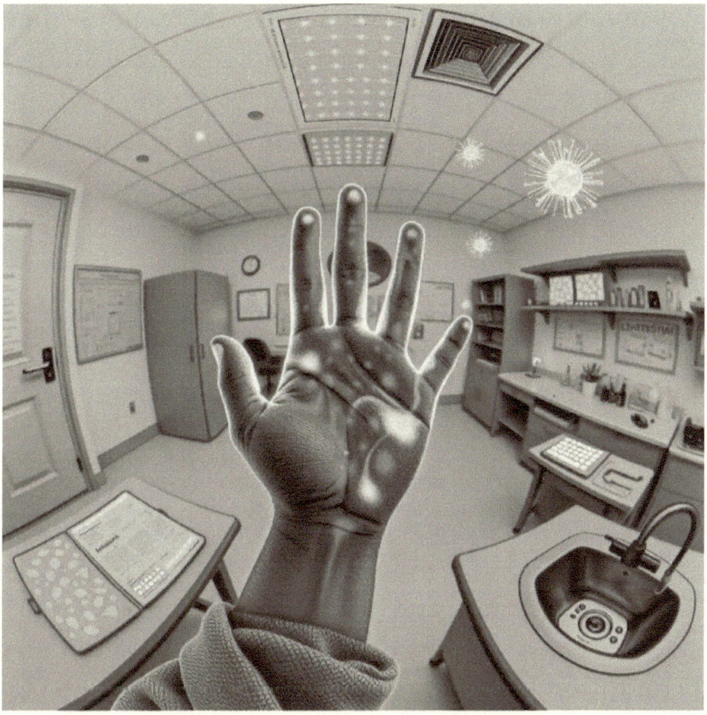

Another project could involve testing handwashing methods—soap vs. water alone, or length of washing time—by measuring bacterial growth afterward. Students may also conduct surveys about hygiene practices in their school and relate those habits to findings from their microbial cultures. These projects not only teach microbiological methods but also instill responsibility and awareness about disease prevention. By becoming detectives in their own classrooms and homes, students learn that science is not confined to a lab—it's alive in the spaces they occupy every day.

Not all science communication needs to be formal or clinical. In fact, creative storytelling is one of the most powerful ways to share microbiological discoveries.

Students can write short stories, comic strips, or poems from the perspective of a microbe. Imagine a virus explaining how it tricks cells, or a helpful gut bacterium talking about keeping humans healthy. Dramatized presentations—like mock interviews with bacteria or news reports on a "mold outbreak"—make microbiology engaging and relatable. Some students might choose to create animated videos, infographics, or podcasts on topics like antibiotic resistance or fungal diseases.

These projects build communication, empathy, and public speaking skills while allowing students to integrate science with language, art, and media. Storytelling in science also fosters a deeper connection—transforming microbes from abstract concepts into characters with roles and stories that young audiences can understand and remember.

Becoming a microbe detective means more than conducting experiments—it means developing a curious eye, an analytical mind, and a creative heart. Through fairs, models, investigations, and stories, students not only learn scientific concepts—they build skills in observation, research, collaboration, and expression. Microbiology offers a world of mysteries waiting to be solved, and youth are perfectly placed to uncover them. This chapter is an open invitation to explore the unseen universe with tools as simple as a notebook, a microscope, or a lump of clay. The next breakthrough in microbial science may not come from a professional lab—but from a classroom filled with young explorers ready to ask the right questions.

End of the Chapter

Chapter-20.
Microbiology Careers – From Lab Coats to Space Suits

Discover how a love for microbes can become a life-changing career. This chapter explores inspiring paths from medicine to space science, showing how curiosity, skill, and passion can turn microbiology into a powerful force for shaping the future.

Microbiology isn't just a subject—it's a gateway to an entire universe of exciting careers that can change lives, improve the planet, and push the boundaries of science. For young learners curious about how to turn their passion for science into a profession, this chapter opens the door to real opportunities. From saving lives in hospitals and tracking global pandemics to creating eco-friendly products and even helping space agencies plan life on Mars, microbiology offers careers that are as diverse as they are impactful. This chapter dives deep into the wide variety of roles available, the skills required, and the inspiring real-world stories of people who turned curiosity into meaningful work. For students standing at the crossroads of choosing their future, microbiology is not just a career path—it's a calling for those ready to make a difference.

The first step in exploring a microbiology career is understanding who the microbe experts are. A microbiologist is a scientist who studies microorganisms in general—including bacteria, viruses, fungi, protozoa, and algae. Their work ranges from basic research in laboratories to applied research in industries like agriculture, pharmaceuticals, and environmental science. Some specialize further into roles like a virologist, who focuses on viruses and viral diseases—studying everything from the flu to emerging threats like Ebola and COVID-19. Virologists help develop vaccines, model outbreaks, and explore how viruses evolve. Another crucial career path is that of an epidemiologist—public health detectives who analyze patterns of disease spread in populations. During pandemics, epidemiologists are the people tracking infections, identifying hotspots, and recommending health strategies. Each of these roles requires keen scientific insight, collaboration, and a desire to solve problems that directly affect human well-being. These scientists are not hidden behind test tubes—they are on the frontlines of discovery, policy, and progress.

Beyond the lab, microbiology opens doors to industries that impact everyday life. Biotechnology companies rely on microbiologists to engineer microbes for products like vaccines, biofuels, and even sustainable packaging. In food safety, professionals ensure the cleanliness and nutritional quality of what we eat—testing for harmful bacteria in dairy, meats, and canned goods. Public health specialists use microbial data to track antibiotic resistance, plan vaccination campaigns, and monitor water and sanitation systems in cities and villages alike. Then there's the exciting field of forensic microbiology, where microbes left behind on a crime scene—on skin, clothing, or even soil—can help solve criminal cases, much like fingerprints. Microbiologists are also vital in agriculture (as soil and plant specialists), pharmaceuticals (in drug development), and even the beauty industry (studying skin and gut flora). These roles may sound different, but they all share one thing: a foundation in microbiological knowledge applied to solving real-world challenges with science, innovation, and care. To succeed in a microbiology-related career, students need more than just academic knowledge. The most valuable tools are often the ones we carry within: curiosity to ask the right questions, observation to notice the details that others miss, and critical thinking to analyze data, spot patterns, and solve problems creatively. Microbiology also teaches resilience—experiments may fail, data may surprise you, and microbes often don't behave as expected. That's when persistence and adaptability become essential. Modern microbiologists must also be comfortable with technology and collaboration, as the field now integrates computer modeling, artificial intelligence, and interdisciplinary teamwork.

For students, developing skills in lab work, data analysis, writing, and communication is just as important as memorizing scientific facts. Joining science clubs, participating in fairs, watching documentaries, and engaging in online microbiology courses are excellent ways to start preparing early. These soft and hard skills combined form the backbone of a strong scientific career.

Real-life stories offer inspiration that theory never can. Take Dr. Rita Colwell, one of the world's leading environmental microbiologists, who helped pioneer the understanding of how climate change affects the spread of diseases like cholera. Or Dr. Kiran Mazumdar-Shaw, India's biotech trailblazer, who began as a fermentation scientist and went on to found one of Asia's leading biopharmaceutical companies, Biocon. Then there's Dr. Anthony Fauci, whose expertise in immunology and infectious diseases guided millions through public health crises. Their journeys show that microbiology can lead to innovation, leadership, and social impact. For students mapping their own paths, it's important to set goals, stay curious, seek mentors, and be open to unexpected turns. Some might start in one field and discover their passion in another—perhaps beginning in plant microbiology and ending up working with NASA on space agriculture. Career planning should be flexible and fueled by genuine interest and a desire to explore the unknown.

In a world increasingly shaped by invisible forces—pandemics, antibiotic resistance, food insecurity, and environmental degradation—microbiologists are becoming the heroes of modern science. Their work quietly supports our health, safety, and survival. For youth choosing their path in life, microbiology offers not only stability and innovation but also the rare opportunity to contribute meaningfully to the future of humanity and the planet. Whether in a white lab coat or a space suit, the world of microbes awaits those bold enough to ask the right questions and prepared enough to chase the answers.

The End

www.ingramcontent.com/pod-product-compliance
Lightning Source LLC
LaVergne TN
LVHW100733061126
851857LV00011B/510
* 9 7 8 9 3 7 2 4 4 6 1 5 9 *